Advances in Cholesteric
Liquid Crystals

Advances in Cholesteric Liquid Crystals

Special Issue Editor
Michel Mitov

MDPI • Basel • Beijing • Wuhan • Barcelona • Belgrade

Special Issue Editor
Michel Mitov
Centre National de la Recherche Scientifique
France

Editorial Office
MDPI
St. Alban-Anlage 66
4052 Basel, Switzerland

This is a reprint of articles from the Special Issue published online in the open access journal *Crystals* (ISSN 2073-4352) in 2019 (available at: https://www.mdpi.com/journal/crystals/special_issues/Cholesteric_Liquid).

For citation purposes, cite each article independently as indicated on the article page online and as indicated below:

LastName, A.A.; LastName, B.B.; LastName, C.C. Article Title. *Journal Name* **Year**, *Article Number*, Page Range.

ISBN 978-3-03928-228-9 (Pbk)
ISBN 978-3-03928-229-6 (PDF)

Cover image courtesy of Michel Mitov.

© 2020 by the authors. Articles in this book are Open Access and distributed under the Creative Commons Attribution (CC BY) license, which allows users to download, copy and build upon published articles, as long as the author and publisher are properly credited, which ensures maximum dissemination and a wider impact of our publications.
The book as a whole is distributed by MDPI under the terms and conditions of the Creative Commons license CC BY-NC-ND.

Contents

About the Special Issue Editor . . . . **vii**

Preface to "Advances in Cholesteric Liquid Crystals" . . . . **ix**

Yuqi Han, Yan Jiang and Wei Guo
Sensing Characteristics of Side-Polished Fiber Based on the Alterations in Helical Structure of Thermo-Sensitive Cholesteric Liquid Crystals
Reprinted from: *Crystals* **2019**, *9*, 465, doi:10.3390/cryst9090465 . . . . **1**

Yueda Liu, Yan Li, Quanming Chen, Sida Li and Yikai Su
Liquid Crystal Based Head-Up Display with Electrically Controlled Contrast Ratio
Reprinted from: *Crystals* **2019**, *9*, 311, doi:10.3390/cryst9060311 . . . . **11**

Chen-Yue Li, Xiao Wang, Xiao Liang, Jian Sun, Chun-Xin Li, Shuai-Feng Zhang, Lan-Ying Zhang, Hai-Quan Zhang and Huai Yang
Electro-Optical Properties of a Polymer Dispersed and Stabilized Cholesteric Liquid Crystals System Constructed by a Stepwise UV-Initiated Radical/Cationic Polymerization
Reprinted from: *Crystals* **2019**, *9*, 282, doi:10.3390/cryst9060282 . . . . **20**

Mikhail N. Krakhalev, Rashid G. Bikbaev, Vitaly S. Sutormin, Ivan V. Timofeev and Victor Ya. Zyryanov
Nematic and Cholesteric Liquid Crystal Structures in Cells with Tangential-Conical Boundary Conditions
Reprinted from: *Crystals* **2019**, *9*, 249, doi:10.3390/cryst9050249 . . . . **28**

Ziheng Wang, Pardis Rofouie and Alejandro D. Rey
Surface Anchoring Effects on the Formation of Two-Wavelength Surface Patterns in Chiral Liquid Crystals
Reprinted from: *Crystals* **2019**, *9*, 190, doi:10.3390/cryst9040190 . . . . **40**

Chia-Hua Yu, Po-Chang Wu and Wei Lee
Electro-Thermal Formation of Uniform Lying Helix Alignment in a Cholesteric Liquid Crystal Cell
Reprinted from: *Crystals* **2019**, *9*, 183, doi:10.3390/cryst9040183 . . . . **62**

Henricus H. Wensink
Effect of Size Polydispersity on the Pitch of Nanorod Cholesterics
Reprinted from: *Crystals* **2019**, *9*, 143, doi:10.3390/cryst9030143 . . . . **72**

Yuri Yevdokimov, Sergey Skuridin, Viktor Salyanov, Sergey Semenov and Efim Kats
Liquid-Crystalline Dispersions of Double-Stranded DNA
Reprinted from: *Crystals* **2019**, *9*, 162, doi:10.3390/cryst9030162 . . . . **83**

About the Special Issue Editor

Michel Mitov is Director of Research at CNRS and Principal Investigator of the Liquid Crystals theme at Centre d'Elaboration de Matériaux et d'Etudes Structurales (CEMES) in Toulouse. His current interests lie in the design and optical properties of complex cholesteric liquid crystal structures (pitch gradient, double helicity, spatially variable helical axis), and biomimetic materials inspired from insect carapaces. He is the inventor of patents related to smart reflective windows to control solar light and heat. He has published an essay of science popularization on soft matter (Sensitive Matter, Harvard University Press), and a textbook on liquid crystal science (Les cristaux liquides, Presses Universitaires de France).

Preface to "Advances in Cholesteric Liquid Crystals"

Most of the optical properties of liquid crystals (LCs) are due to their chiral structures. With their helical structure, cholesteric LCs (CLCs)—also named chiral nematic LCs—figure prominently in LC science. The selective reflection of light is their flagship property. A properly oriented layer of CLC is a unique multifunctional material: It is, at the same time, reflector, notch filter, polarizer, and optical rotator.

The research on CLCs often requires making a continuous description of chiral structures and their properties from the nanometer range to the macroscopic scale. It provides natural links between different disciplines in soft matter science and well beyond. It promises a myriad of applications in the areas of sensors based on color changes, tunable bandpass filters, rewritable color recordings, hyperspectral imaging, thermal printable e-paper, polarizer-free reflective displays, lasing, microlenses, smart windows, and many more. Lyotropic CLC formulations may be chosen for their beauty, to screen UV light, or to obtain iridescent visual effects. The CLC structure has very far-reaching implications in technology which have not yet shown their full significance.

The CLC structure is a ubiquitous design in the animal and plant kingdoms. The functions of biological CLCs include maximization of packing efficiency, optical information, radiation protection, and mechanical stability. The role of the cholesteric structure in living matter is far from being fully defined, and many biomimetic replicas are still awaiting realization. This book is a series of papers written by researchers working in LC science and technology, and covers experimental investigations, theoretical aspects, and applications. It includes one review paper on the packing of double stranded DNA molecules [Y. Yevdokimov et al.] and seven research articles that may fall into the following two groups:

Fundamental investigations of defects, textures, and structures:

Experimental studies and computer simulations of director configurations and topological defects in nematic and cholesteric layers with tangential–conical boundary conditions [M. N. Krakhalev et al.].

Theoretical investigations about the role of length polydispersity on the helical pitch of nanorod based CLCs with continuous length polydispersity [H. H. Wensink].

Mechanisms of voltage-induced changes of CLC textures by dielectric heating in the case of pulses shaped to align the uniform lying helix texture [C.-H. Yu et al.].

Theoretical analysis of surface patterns formed at the free surface of CLCs with a focus on the multiple-length-scale surface wrinkling phenomenon [Z. Wang et al.].

Experimental studies aimed at applications:

Optical fibers coated with a CLC as temperature sensors [Y. Han et al.].

Head-up displays for improving automobile driving safety based on the dual role of reflector and polarizer of CLCs [Y. Liu et al.].

CLC materials for field-induced opaque-to-transparent smart windows made of a hybrid association of polymer-dispersed and polymer-stabilized LC morphologies [C.-Y. Li et al.].

I hope that reading this set of contributions proves to be an enjoyable and illuminating experience.

Michel Mitov
Special Issue Editor

Article

Sensing Characteristics of Side-Polished Fiber Based on the Alterations in Helical Structure of Thermo-Sensitive Cholesteric Liquid Crystals

Yuqi Han [1,2,*], Yan Jiang [1] and Wei Guo [3]

1. Guangdong Vocational College of Post and Telecom, Guangzhou 510630, China
2. School of Network and Continuing Education, Beijing University of Posts and Telecommunications, Beijing 510630, China
3. China Tower Corporation Limited Guangdong Branch, Guangzhou 510000, China
* Correspondence: hanyuqi8373@163.com or hanyuqi@gupt.net

Received: 19 July 2019; Accepted: 1 September 2019; Published: 5 September 2019

Abstract: Cholesteric liquid crystals (CLCs) are sensitive to environmental temperature changes, and have been employed as a specific intermediary for biosensors. Considering the temperature-dependent structural changes of CLCs, this study aimed to determine the sensing properties of side-polished fibers (SPFs) after coating with CLCs. The experimental results demonstrated that, with regard to the transmitted spectrum, the loss peak of CLC-coated SPFs exhibited a positive linear relationship with temperature changes over a range of 20 to 50 °C. The linear correlation coefficient achieved 97.8% when the temperature increased by 10 °C, and the loss peak drifted by 12.72 nm. The reflectance spectrum of CLCs coated on the polished surface were obtained using optical fiber sensors. The feasibility of measuring the helical structure of CLCs was further verified using SPF transmission spectroscopy. The findings indicated that the transmitted spectrum of SPFs could be adopted to characterize the helical structure of CLCs, which lays a solid foundation for further study on SPF-based biosensors.

Keywords: liquid crystal; cholesteric liquid crystals; optical fiber; side-polished fiber; sensor

1. Introduction

Cholesteric liquid crystals (CLCs) with orientational order are capable of undergoing liquid to solid phase transition. They exhibit both liquid fluidity and crystal anisotropy, which lie primarily in optical anisotropy, such as birefringence. In addition to optical anisotropy, CLCs display a selective reflection of light due to their periodic helical structure. The selective reflection properties of CLCs have been shown to be associated with their helical pattern. The intrinsic pitch of the thermos-sensitive CLC selected in this study was the same as the wavelength of visible light. When the spiral axis of CLC is perpendicular to the substrate surface, the Bragg reflection of visible light can appear. The helical structure of CLCs is not only related to its material properties but is also a sensitive function of temperature, electromagnetic field, acoustic field, radiation field, and even biochemical sensing. Therefore, CLCs have been widely applied in sensor fields, especially biosensors [1–7]. The liquid crystalline state has become ubiquitously recognized in organisms, in which the helical structure of amphiphilic lipids from the plasma membrane is a natural composition containing a lyotropic liquid crystal structure. Collagen and organismal DNA may synthesize CLCs with a helical or double helical structure through self-assembly [8]. However, the most commonly employed methods for characterizing CLCs in biosensors are polarized light microscopy [8], differential scanning calorimetry, X-ray diffraction, atomic force microscopy, Fourier-transform infrared spectroscopy, circular dichroism spectroscopy, nuclear magnetic resonance, and small-angle neutron scattering [8]. By examining the transmittance and texture alterations of liquid crystal molecules under a polarizing microscope, the phase transition of CLCs can be analyzed, as well as the morphological variables, such as changes in texture and orientation. However, in view of practical applications, it is rather difficult to accurately and comprehensively characterize the liquid crystal state, regardless of the unavailability of miniaturized analytical equipment. Thus, inevitably, the quantification of the measurement data must be done focally by using micro-point probes, rather than real-time monitoring of changes in CLCs [9–11].

Optical fiber sensors, especially those with evanescent wave absorption, have already been applied in the field of biomedical sensing research [12–14]. Side-polished fiber (SPF) is a specific optical fiber made by removing part of the cladding from standard optical fibers through optical microfabrication technology [15–17]. A variety of new optical devices and sensors have been fabricated by coating with liquid crystals on their polished surfaces. Several studies have [18,19] proposed that azobenzene polymer can be introduced to the nematic liquid crystal and coated on the polished surface of SPF, through which an optically controllable SPF attenuator is made based on the liquid crystals. However, Yu et al. [20,21] coated the mixed liquid crystals onto the side-polished surface of SPF and managed to fabricate an optically powered fiber optic sensor. Tang et al. [22] successfully fabricated optical fiber-based volatile organic compound gas sensors by coating CLCs onto the polished surface of SPFs. In addition to the advantages of small volume, low cost, anti-electromagnetic interference, and being easily accessible and reusable, SPFs are sensitive to external environmental factors. This is probably because the optical field in the core leaks out through the polished surface in the form of an evanescent wave, which makes it possible to interact with the external environmental factors. Therefore, SPF is not only sensitive to the external environmental changes but also has adjustable length and depth. SPFs often exhibit longer operating distances or larger surfaces, and thus they tend to have relatively higher sensitivity. According to the findings of previous experiments, we hypothesize that the orientational changes in nematic liquid crystals can be characterized by SPF-transmitted optical power [23].

In this study, the sensing properties of SPFs based on the helical structure changes of thermos-sensitive CLCs were investigated within the visible light range. The reflection spectra of thermos-sensitive CLCs coated upon the polished surface were measured using an optical fiber reflection probe, while the transmitted spectra of SPFs were used to characterize the temperature-induced structural changes of CLCs. The findings provide the feasibility of measuring the helical structure changes of CLCs by SPF via transmission spectroscopy, as well as laying a foundation for future research on CLC-coated SPF biosensor.

2. Materials and Methods

2.1. Experimental Design

CLCs has been recognized as chiral nematic liquid crystals. Chirality refers to the ability to selectively reflect a certain portion of circular polarization. The CLC molecules exhibit a uniform alignment of their direction in a single layer and accumulated layer by layer. The directional vectors of each layer are slightly deviated from those of the adjacent layer, thereby forming a helical structure. The helical axis of CLC is perpendicular to $\overline{n}$, which corresponds to the optical axis. Under a 360° rotation along the helical direction, its long axis returns to the initial orientation. The pitch (periodic interval between layers) of CLC is an important parameter for determining its helical structure. CLCs demonstrate the characteristics of reflecting visible light with one pitch, which is strongly dependent on temperature changes and external environmental factors.

Hence, the experimental design of this study was focused on CLCs (5 µL) coated on the polished surface of SPFs. The intrinsic pitch of cholesterol ester liquid crystal selected in this study was equivalent to the wavelength of visible light. The helical structure of cholesterol ester liquid crystals may change in response to increasing temperatures, and disappear upon reaching isotropy [24–26]. Considering the temperature-sensitive structural changes of CLCs, the sensing response characteristics of SPFs were evaluated. Through optical fiber sensing, the ambient temperature-induced alterations in the reflective spectrum of CLCs were determined, and the relationship between its reflective spectrum and the helical structure was examined. Moreover, the characteristics of the transmission spectrum of SPFs in response to the ambient temperature-induced structural changes of CLCs were examined. Additionally, the transmission spectrum of SPFs was used to verify the feasibility of measuring the structural changes of CLCs.

2.2. Preparation of CLCs

CLCs are sensitive to the changes in environmental temperature and are often displayed as bright colors. In addition, they are relatively similar to proteins with respect to the spatial uniformity of their helical structures. The temperature sensitivity range of liquid crystals can be adjusted by mixing different types of formulations. This experimental design was based on the studies of Brown et al. [24] and Elser et al. [25]. Specifically, cholesteryl oleyl carbonate (15, 115-7; Sigma-Aldrich), cholesteryl pelargonate (C7, 880-1; Sigma-Aldrich), and cholesteryl benzoate (C7, 580-2; Sigma-Aldrich) were used to prepare the temperature-sensitive cholesterol ester liquid crystals. The preparation procedure was as follows. First, 0.38, 0.52, and 0.10 g of cholesterol carbonate, cholesterol nonyl ester, and cholesterol benzoate, were, respectively, mixed in a glass bottle. Next, the tube bottle was heated by a hot air cylinder until all the powdered solid samples were melted into a transparent liquid. Subsequently, the melted liquid sample was placed in an ultrasonic cleaner, and uniformly homogenized at 40 °C for 10 minutes. After ultrasound homogenization, the sample was transferred onto a tube mixer and oscillated for 10 minutes, to completely mix it again. Thereafter, the sample was statically cooled at room temperature, and the visualized color was recorded (Figure 1).

Figure 1. The preparation of cholesteric liquid crystals (CLCs).

2.3. Thermosensitivity of CLCs

The homochiral helical structure of CLCs allows the Bragg reflection of visible light when the helical axis of the liquid crystals is perpendicular to the substrate (Figure 2). Typically, CLCs can exert the strongest selective reflection on the wavelengths of the incident light only if the light is incident along the spiral axis of liquid crystals. The formula for reflection wavelength (λ_0) is shown as follows:

$$\lambda_0 = \bar{n}p \tag{1}$$

where, $\bar{n}$ is the equivalent refractive index of the CLC. Thus, the selective reflection properties of CLCs are related to their helical structures.

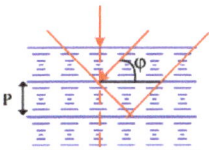

Figure 2. CLCs exhibit the strongest selective reflection of incident light wavelength, especially when the incident light is projecting along its spiral axis.

The helical structures of CLCs can be altered in response to high temperature [25]. According to Formula (1), the reflection wavelength λ_0 could drift, while the colors of reflected light varied with the increasing temperatures (Figure 3b). The liquid crystalline form of cholesteryl ester was blue–green in color at room temperature (~20 °C), but turned green–yellow, after being heated to 50 °C. The liquid crystal of cholesteryl ester transited from the anisotropic phase to the isotropic phase. Specifically, its spiral structure disappeared after being heated to 70 °C, and it achieved isotropy and displayed as a transparent liquid. The helical structure of CLCs can be confirmed by the wavelength-selective reflection from CLCs. The wavelength-selective reflection arises from the Bragg reflection with the periodic helical structure in CLCs. According to the Bragg reflection, when the incident light illuminates the CLC film vertically, the light at the wavelength λ_0 will be reflected while the light at other wavelength transmits [18]. Therefore, the helical structure of CLCs may react to environmental temperature changes and appears to be quite sensitive.

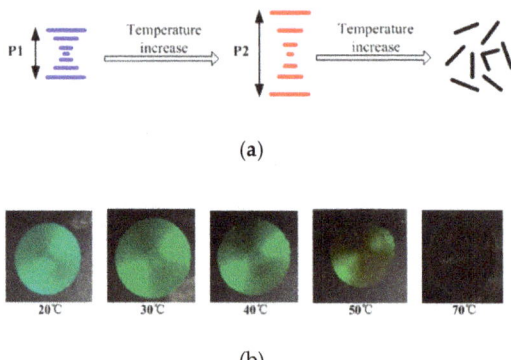

Figure 3. The effect of temperature on the helical structure of CLCs. (**a**) The schematic diagram for the structural changes of CLCs during increasing temperatures. (**b**) The experimental color photographs of the reflected light of CLCs observed by the naked eye from 20 to 70 °C.

2.4. Preparation of SPFs

In the present study, the wheel side-polishing technique was employed for the preparation of single-mode optical fibers (SMF-28e single-mode ordinary communication optical fibers, Corning Company) [16]. The SPFs used in this experiment were prepared by polishing twice. The transmission power of SPFs was 2.2 dBm before polishing, and −2.3 dBm after polishing. The residual thickness of the SPF was obtained by measuring the side surface of the SPF using a microscope at 10× magnification. During microscopic measurements, the mechanistic platform of the microscope was moved along the axis of SPFs in steps of 1.0 mm, and a series of images were taken at different positions of SPFs. As shown in Figure 4a,c are transitional zones at both ends of the SPF, while (b) is a flat zone in the middle. The length of the polished surface and average residual thickness of the SPF were found to be ~20 mm and ~2 μm, respectively (Figure 5).

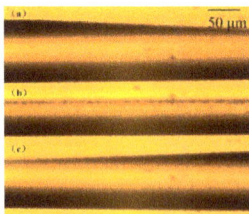

Figure 4. Side view of side-polished fibers (SPFs). (**a**) and (**c**) are the transition areas, while (**b**) is the flat zone of the SPF.

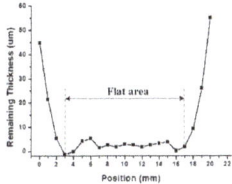

Figure 5. The length of the polished surface and residual cladding thickness of SPFs.

2.5. Reflection Spectrum Measurement of CLCs

The CLC layer coated on the surface of SPFs was regarded as the Bragg reflection structure with liquid crystal pitch as its period. The maximum reflection wavelength was calculated according to the following formula:

$$\lambda_0 = 2n_{CLC}\Lambda = 2n_{CLC}p, \qquad (2)$$

where n_{CLC} is the effective refractive index of the cladding of CLC under Bragg reflection waveguide structure, Λ is the time period of Bragg waveguide, and p is the pitch of CLC. Given that CLCs are temperature sensitive, its helical structures could be altered in response to increasing temperatures. Therefore, raising the temperature might shift the maximum reflection wavelength satisfying the Bragg reflection.

2.6. Transmission Spectrum Measurement of SPFs

The SPF used in our study was made from a single mode fiber using the wheel-polishing method. In a step-index fiber with a circular cross-section, the refractive index of a fiber's core, n_{co}, is greater than that of its cladding, n_{cl}, and the fiber mode effective index, n_{eff}, lies between n_{co} and n_{cl}. In the case of the SPF having a D-shaped cross-section, the cladding between the core and the polished surface can be made to be sufficiently thin to allow the evanescent field to extend into the external medium above the polished surface of the SPF. If the polished surface of the SPF is coated with a material with refractive index, n_{CLC}, the mode effective index of the SPF/external material medium structure will depend on the cladding thickness and the value of the n_{CLC}. According to the theoretical analysis on optical propagation characteristics of SPF shown in reference [15] and [16], when $n_{CLC} < n_{eff}$, there is total internal reflection and the loss of light propagating in SPF is the minimum. When $n_{CLC} \approx n_{eff}$, the fiber mode becomes leaky, and power can be radiated out of the core of the fiber. On this occasion, the loss of light propagating in the SPF is the maximum due to the guided mode changing to radiation mode. When $n_{CLC} > n_{eff}$, there is semi-reflection and part of light leaks. The loss of light propagating in an SPF can, therefore, be very dependent on the value of the n_{CLC}.

The transmitted light on the SPF was incident to the Bragg reflection from the CLC at an angle of θ. Because the evanescent field leaked from the SPF core, the light wave from the CLC was reflected back to the side-polished optical fiber for the purpose of light propagation (Figure 6). Therefore, the CLC coated on the polished surface of SPFs might constitute liquid–crystal waveguides cladding with a Bragg reflection structure, and its reflective wavelength was calculated as follows:

$$m\lambda_0 = 2n_{CLC}\, p\cos\theta \qquad m = 0,1,2,3... \qquad (3)$$

Considering that CLCs are temperature sensitive, their spiral structure may change in response to increasing temperatures. It was speculated that the spiral axis of CLCs was perpendicular to the polished surface of SPFs, and therefore, the liquid crystal molecules were intrinsically twisted along the polished surface to form periodic helical structures (Figure 6). In addition, the reflected light waves satisfying the Bragg reflection waveguide were propagated from CLCs to SPFs, and thus the transmitted spectrum of CLC-coated SPFs drifted with the increase in temperature. Furthermore, the reflective spectra of CLCs and the transmitted spectra of SPFs were measured simultaneously, and the results were compared.

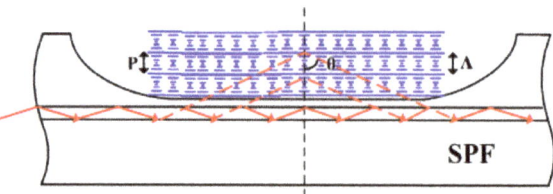

Figure 6. Structure of CLC-coated SPFs and schematic illustration of its temperature sensing.

2.7. Experimental Device

The schematic diagram of the experimental device for sensing the characteristics of SPFs based on the structural changes of CLCs is illustrated in Figure 7. LS-1 provides detection light covering the visible band. The transmitted spectra of SPFs coated with CLCs on its polished surface, and the reflection spectra of liquid crystals were measured by using the transmission probe of USB4000 Spectrometer (Ocean Optics) and the optical fiber reflection probe perpendicular to CLCs. A hot plate was used to control the temperature around the polished surface of SPFs, by heating from room temperature 20 to 70 °C. Before heating, the liquid crystal was transformed into a transparent liquid, while the heat disc was adjusted to the targeted temperature. Such a process is exploited here to avoid the inhomogeneity caused by daubing and is capable of making CLCs reorient and self-assemble in order on the polished surfaces of SPFs, and thus form a more stable and homogeneous helical microstructure in the CLCs layer.

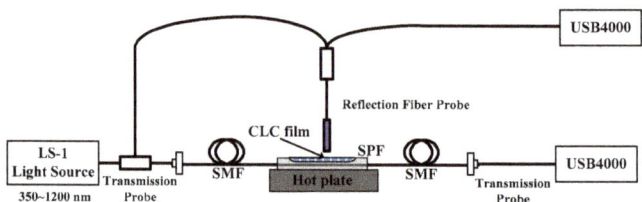

Figure 7. Schematic diagram of the experimental device for evaluating the sensing characteristics of CLC-coated SPFs.

3. Results and Discussion

CLCs have a selective reflection of light, resulting from the fact that they are composed of a periodic helical structure. In this study, temperature-sensitive CLCs were uniformly coated on the polished surface of SPFs, and the sensing characteristics of SPFs were evaluated based on the structural change of CLCs.

3.1. Reflection Spectra of CLCs in Response to Environmental Temperature Changes

The optical fiber reflector probe was placed vertically above the CLCs, and the light reflection was incident vertically on the coated CLC surface. When the ambient temperature around the polished surface of SPFs was increased from 20 to 50 °C at intervals of 10 °C, the maximum reflection wavelength of the CLC reflectance spectrum was shifted from 536 to 587 nm. When the temperature was raised to 70 °C, the reflection peak of CLCs disappeared (Figure 8). Formula (2) showed that the optical selective reflection properties of CLCs are related to their helical structures. The helical structures of CLCs altered with the temperature rising from 20 to 50 °C. The reflection wavelength measured by optical fiber reflector was red-shifted by approximately 50 nm. CLCs reached isotropy at the phase transition temperature of approximately 70 °C, and their periodic helical structure did not exist anymore. Within the temperature range of 20 to 50 °C, the maximum reflection wavelength of CLCs was linearly correlated with the drift of ambient temperature. The linear equation of λp = 1.7341T + 497.69 was obtained, and the linear correlation coefficient achieved 94.5% (Figure 9). These results indicated that the reflectance spectra of CLCs exhibit a good linear relationship with temperature. Notably, the maximum reflection wavelength of CLC reflectance spectra shifted 17.34 nm for every 10 °C rise in temperature. Therefore, the optical selective reflection properties of CLCs strongly support that its helical structure has a temperature-sensitive function.

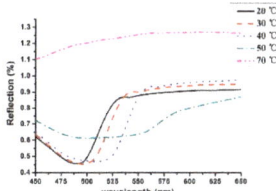

Figure 8. Temperature dependence of CLC reflectance spectra measured by an optical fiber sensor.

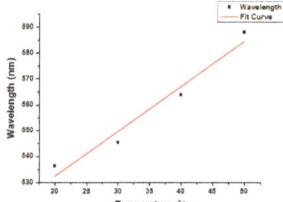

Figure 9. Linear fitting curves for the maximum reflection wavelength of CLCs over a temperature drift.

For the typical case where the helix axis is perpendicular to the interface, it can be seen from Formula (2) that the measured reflection spectrum of the CLC surface may display a significant reflection peak (Figure 10a). However, in this experiment, the spiral axis of CLCs around the polished surface is not uniformly perpendicular to the interface due to the coating of CLCs on the SPF polished surface (Figure 10b). As a result, the Bragg's reflection peaks of the CLC surface were wide, and only the corresponding wavelength at the edge of the reflection peaks was affected by temperature alterations.

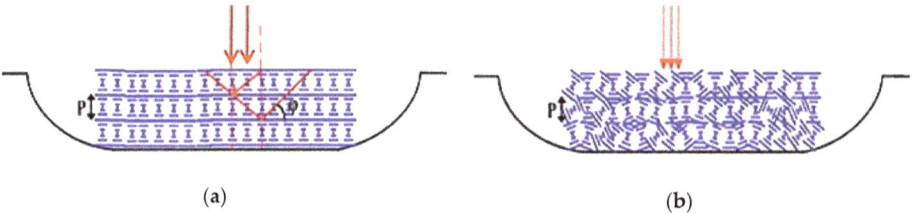

Figure 10. Schematic diagram for the helical arrangement of CLCs coated on the SPF polished surface. (a) The helix axis of CLCs coated on the SPF is perpendicular to the polished surface ideally. (b) The helix axis of CLCs coated on the SPF is not uniformly perpendicular to the interface in the experiment.

3.2. Transmission Spectra of SPFs in Response to the Structural Changes of CLCs

The transmission spectra of CLC-coated SPFs are presented in Figure 11 as a function of temperature. The transmission light in the SPFs was incident into the Bragg reflection waveguide structure composed of CLC, and the light waves satisfying the Bragg reflection could be reflected back from CLCs to SPFs to maintain propagation. To compare the transmission spectra of CLC-coated SPFs and those of CLCs alone, the loss peak (valley) of SPF transmission spectra was assessed in the present study. When the temperature was increased from 20 to 70 °C, the transmission spectra of CLC-coated SPFs blue-shifted from 702.77 to 597.8 nm, suggesting that the helical structure of CLCs changed and the long-wave band of visible light was transmitted back to SPFs in response to higher temperatures. These results were consistent with the red-shifted spectral data of CLCs (Figure 8). This indicates that the transmission spectra of SPFs can characterize the structural changes of CLCs. When the temperature reached 70 °C, the helical structure of CLCs disappeared, indicating the phase transition from liquid crystal to isotropy. The measured transmission spectrum of SPFs was different between each temperature, at which the half-width of loss transmission peak was the largest. Therefore, the transmitted spectra of SPFs can characterize the changes in the helical structure of CLCs modulated by ambient temperature.

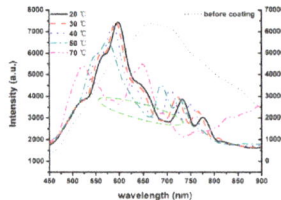

Figure 11. Temperature dependence of loss peaks pertaining to the transmission spectra of CLC-coated SPFs.

Where the helix axis is perpendicular to the interface (Figure 10a), the transmission spectra of CLC-coated SPFs may display a significant loss peak. However, in this experiment, without the alignment of CLC, the spiral axis of CLCs around the polished surface was not uniformly perpendicular to the interface due to the coating of CLCs on SPF polished surface (Figure 10b). As a result, the Bragg's reflection peaks of CLC surface were wide, and only the corresponding wavelength at the edge of the reflection peaks was affected by temperature alterations. Thus, the cause of many small peaks near the main peak shown as Figure 11 should be the spiral axis of CLCs adjacent to the polished surface, which was not uniformly perpendicular to the interface.

The CLCs coated on the polished surface of SPFs have a periodic helical structure within the temperature range of 20 to 50 °C. The transmitted spectrum loss peak of SPFs is based on the fact that changes in the helical structure of CLCs are a temperature-sensitive function. The linear equation $\lambda p = 1.272T + 729.891$ was fitted and its linear correlation reached 97.8% (Figure 12). The results indicate that the transmitted spectrum loss peak of CLC-coated SPFs exhibits a good linear relationship with temperature. The blue-shifted transmission loss peak of SPFs was 12.72 nm for every 10 °C rise in temperature. Therefore, within the temperature range of 20 to 50 °C, SPFs, based on the spiral structure changes of CLC, can sense the alterations in environmental temperature, with relatively higher sensitivity and better linear relationship.

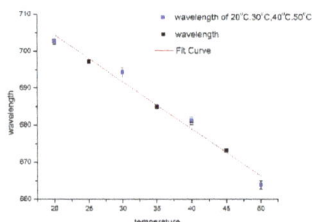

Figure 12. Linear fitting curves for loss peaks pertaining to the transmission spectra of CLC-coated SPFs over a temperature drift.

4. Conclusions

In conclusions, the sensing properties of CLC-coated SPFs in response to the ambient temperature-modulated structural change of CLCs were assessed in the present study. The loss peak of the transmitted spectrum pertaining to CLC-coated SPFs demonstrated a good linear relationship with temperature (ranging from 20 to 50 °C). The linear correlation coefficient achieved 97.8% when the temperature increases by 10 °C, and the loss peak drifted by 12.72 nm. At 70 °C, the half-maximum width of the loss transmission peak of CLCs reached the maximum value due to the transition from the liquid crystal phase to the isotropic phase. The maximum reflection wavelength of CLCs red-shifted 17.34 nm for every 10 °C rise in temperature, and the linear correlation coefficient achieved 94.5% at the temperature range of 20 to 50 °C. When the temperature rose to 70 °C, the reflection peak of CLCs disappeared. The optical selective reflection characteristics of CLCs well-demonstrate that its helical structure is a temperature-sensitive function, as evaluated by the transmitted spectrum analysis of SPFs. The feasibility of measuring the helical structure of CLCs by SPF transmission spectra was further verified by the reflectance spectrum of CLCs assessed by a fiber-optic reflection probe. Collectively, the transmission spectra of SPFs could be used to characterize the structural changes of CLCs, thereby laying a foundation for further investigations on CLC-coated SPF biosensors.

Author Contributions: Conceptualization, Y.H.; methodology, Y.H.; formal analysis, Y.H.; investigation, Y.H.; resources, W.G.; data curation, Y.J.; writing—original draft preparation, Y.H.; writing—review and editing, Y.H.; project administration, Y.H.

Funding: This research was funded by Fundamental Research Funds for Science and Technology Program of Guangzhou, China, grant number 201707010496; Fundamental Research Funds for Youth Innovation Personnel Training Project of Guangdong, China, grant number 2017GkQNCX042.

Acknowledgments: The authors would like to express their gratitude to EditSprings (https://www.editsprings.com/) for the expert linguistic services provided. And the authors would like to express their gratitude to Zhe Chen, Jinan University for the technical support.

Conflicts of Interest: The authors declare no conflict of interest. The funders had no role in the design of the study; in the collection, analyses, or interpretation of data; in the writing of the manuscript, or in the decision to publish the results.

References

1. Woliński, T.R.; Bock, W.J. Cholesteric liquid crystal sensing of high hydrostatic pressure utilizing optical fibers. *Mol. Cryst. Liq. Cryst.* **1991**, *199*, 7–17. [CrossRef]
2. Shibaev, P.V.; Schlesier, C. Distant mechanical sensors based on cholesteric liquid crystals. *Appl. Phys. Lett.* **2012**, *101*, 193503. [CrossRef]
3. Moreira, M.F.; Carvalho, I.C.S.; Cao, W.; Bailey, C.; Taheri, B.; Palffy-Muhoray, P. Cholesteric liquid-crystal laser as an optic fiber-based temperature sensor. *Appl. Phys. Lett.* **2004**, *85*, 2691–2693. [CrossRef]
4. Mitov, M. Cholesteric Liquid Crystals with a Broad Light Reflection Band. *Adv. Mater.* **2012**, *24*, 6260–6276. [CrossRef]
5. Hennig, G.; Brittenham, G.M.; Sroka, R.; Kniebühler, G.; Vogeser, M.; Stepp, H. Bandwidth-variable tunable optical filter unit for illumination and spectral imaging systems using thin-film optical band-pass filters. *Rev. Sci. Instrum.* **2013**, *84*, 043113. [CrossRef]
6. Hikmet, R.A.M.; Kemperman, H. Electrically switchable mirrors and optical components made from liquid-crystal gels. *Nature* **1998**, *392*, 476–479. [CrossRef]
7. Woltman, S.J.; Jay, G.D.; Crawford, G.P. Liquid-crystal materials find a new order in biomedical applications. *Nat. Mater.* **2007**, *6*, 929–938. [CrossRef]
8. Hsiao, Y.C.; Sung, Y.C.; Lee, M.J.; Lee, W. Highly sensitive color-indicating and quantitative biosensor based on cholesteric liquid crystal. *Biomed. Opt. Express* **2015**, *6*, 5033–5038. [CrossRef]
9. Popov, N.; Honaker, L.W.; Popova, M.; Usol'tseva, N.; Mann, E.K.; Jákli, A.; Popov, P. Thermotropic Liquid Crystal-Assisted Chemical and Biological Sensors. *Materials* **2017**, *11*, 20. [CrossRef]

10. Brake, J.M.; Mezera, A.D.; Abbott, N.L. Active control of the anchoring of 4′-pentyl-4-cyanobiphenyl (5CB) at an aqueous-liquid crystal interface by using a redox-active ferrocenyl surfactant. *Langmuir* **2003**, *19*, 8629–8637. [CrossRef]
11. McCamley, M.K.; Ravnik, M.; Artenstein, A.W.; Opal, S.M.; Žumer, S.; Crawford, G.P. Detection of alignment changes at the open surface of a confined nematic liquid crystal sensor. *J. Appl. Phys.* **2009**, *105*, 123504. [CrossRef]
12. Huang, H.; Zhai, J.; Ren, B. Fiber-Optic Evanescent Wave Biosensor and Its Application. *ACTA Opi. Sin.* **2003**, *23*, 451–454.
13. Deng, L.; Feng, Y.; Wei, L. The Research of Fiber Optic Evanescent Wave Biosensor. *ACTA Photonica Sin.* **2005**, *34*, 1688–1692.
14. Lin, H.Y.; Tsai, W.H.; Tsao, Y.C.; Sheu, B.C. Side-polished multimode fiber biosensor based on surface plasmon resonance with halogen light. *Appl. Opt.* **2007**, *46*, 800–806. [CrossRef]
15. Chen, Z.; Li, F.; Zhong, J. Side polished fiber and application. In Proceedings of the 12th Fiber Communication and 13th Integrated Optics Conference, Jinan University, Guangzhou, China, 2005; pp. 407–412.
16. Chen, Z.; Cui, F.; Zeng, Y. Theoretical analysis on optical propagation characteristics of side-polished fibers. *Acta Photonica Sin.* **2008**, *37*, 918–923.
17. Chen, Z.; Qin, J.; Pan, H.; Zhang, J.; Xiao, Y.; Yu, J. All-fiber integrated optical power monitor-controller. *Chin. J. Lasers* **2010**, *37*, 1047–1052. [CrossRef]
18. Hsiao, V.K.S.; Li, Z.; Chen, Z.; Peng, P.C.; Tang, J. Optically controllable side-polished fiber attenuator with photoresponsive liquid crystal overlay. *Opt. Express* **2009**, *17*, 19988–19995. [CrossRef]
19. Fu, W.H.; Hsiao, V.K.S.; Tang, J.Y.; Wu, M.H.; Chen, Z. All fiber-optic sensing of light using side-polished fiber overlaid with photoresponsive liquid crystals. *Sens. Actuators B Chem.* **2011**, *156*, 423–427. [CrossRef]
20. Yu, J.; Li, X.; Du, Y.; Zhang, J.; Chen, Z. Study of photorefractive properties of liquid crystal hybrid thin film by side polished fiber sensor. In Proceedings of the Third Asia Pacific Optical Sensors Conference, Sydney, Australia, 31 January–3 February 2012; p. 835122.
21. Yu, J.; Li, H.; Hsiao, V.K.; Liu, W.; Tang, J.; Zhai, Y.; Du, Y.; Zhang, J.; Xiao, Y.; Chen, Z. A fiber-optic violet sensor by using the surface grating formed by a photosensitive hybrid liquid crystal film on side-polished fiber. *Meas. Sci. Technol.* **2013**, *24*, 094019. [CrossRef]
22. Tang, J.; Fang, J.; Liang, Y.; Zhang, B.; Luo, Y.; Liu, X.; Li, Z.; Cai, X.; Xian, J.; Lin, H.; et al. All-fiber-optic VOC gas sensor based on side-polished fiber wavelength selectively coupled with cholesteric liquid crystal film. *Sens. Actuators B Chem.* **2018**, *273*, 1816–1826. [CrossRef]
23. Han, Y.; Chen, Z.; Cao, D.; Yu, J.; Li, H.; He, X.; Zhang, J.; Luo, Y.; Lu, H.; Tang, J.; et al. Side-polished fiber as a sensor for the determination of nematic liquid crystal orientation. *Sens. Actuators B Chem.* **2014**, *196*, 663–669. [CrossRef]
24. Stewart, G.T. Liquid Crystals in Biological Systems. *Mol. Cryst. Liq. Cryst.* **1966**, *1*, 563–580. [CrossRef]
25. Elser, W.; Ennulat, R.D. Selective Reflection of Cholesteric Liquid Crystals. *Adv. Liq. Cryst.* **1976**, *2*, 73–172.
26. Zhou, Y.; Huang, Y.; Ge, Z.; Chen, L.-P.; Hong, Q.; Wu, T.X.; Wu, S.-T. Enhanced photonic band edge laser emission in a cholesteric liquid crystal resonator. *Phys. Rev. E* **2006**, *74*, 061705. [CrossRef]

© 2019 by the authors. Licensee MDPI, Basel, Switzerland. This article is an open access article distributed under the terms and conditions of the Creative Commons Attribution (CC BY) license (http://creativecommons.org/licenses/by/4.0/).

Article

Liquid Crystal Based Head-Up Display with Electrically Controlled Contrast Ratio

Yueda Liu, Yan Li *, Quanming Chen, Sida Li and Yikai Su

Department of Electronic Engineering, Shanghai Jiao Tong University, Shanghai 200240, China; Lyueda@sjtu.edu.cn (Y.L.); qm.chen@sjtu.edu.cn (Q.C.); lsd236@sjtu.edu.cn (S.L.); yikaisu@sjtu.edu.cn (Y.S.)
* Correspondence: yan.li@sjtu.edu.cn

Received: 9 May 2019; Accepted: 14 June 2019; Published: 18 June 2019

Abstract: With the growing demand for driving safety and convenience, Head-Up Displays (HUDs) have gained more and more interest in recent years. In this paper, we propose a HUD system with the ability to adjust the relative brightness of ambient light and virtual information light. The key components of the system include a cholesteric liquid crystal (CLC) film, a geometric phase (GP) liquid crystal lens, and a circular polarizer. By controlling the voltage applied to the GP lens, the contrast ratio of the virtual information light to ambient light could be continuously tuned, so that good visibility could always be obtained under different driving conditions.

Keywords: geometry phase; cholesteric liquid crystal; head-up display

1. Introduction

Head-Up Displays (HUDs) have attracted more and more interest recently because of their important role in improving driving safety. A HUD can directly project vital driving information such as speed and navigation into the driver's eyes without the need to shift one's sight off the road. Nowadays, HUDs have been integrated into some top-end vehicles. For example, in BMW M5 (an executive car produced by German automotive manufacturer BMW), a HUD is realized by projecting an image onto a translucent TFT (thin-film transistor) display in the windscreen via a specially shaped mirror [1]. More modern implementations and enhancements of HUDs have also been developed. Okumura et al. realized a wide field-of-view HUD using a partial transparent Fresnel reflector as a combiner [2]. Wei et al. proposed a HUD with superior optical performance using three off-axis freeform mirrors [3]. Zhan et al. used GP elements to enhance the performance of HUDs in many aspects [4].

In different driving conditions, the brightness of ambient light may undergo dramatic change, which would affect the visual effects of HUDs. When a car runs in bright sunlight in summer, the virtual information provided by a HUD gets washed out. Increasing the brightness of the HUD may alleviate the problem, but greatly increases power consumption. When a vehicle runs on a dark road at night, the excessive brightness of the HUD may become quite disturbing. Moreover, sometimes, a sudden change in ambient light may occur when, for instance, entering or exiting a tunnel. However, most current HUDs can only adjust the brightness of the virtual image, but cannot control the brightness of the real world scene. To improve visual quality and driving safety, it is highly desirable to realize real-time brightness adaptation for both ambient light and virtual information, and thus ensure the contrast between the two within a reasonable range.

Several approaches have been proposed to achieve transmission control in augmented reality systems. Photochromic materials [5] have excellent dimming capability, but the response time is too slow for real-time operation. The STN (super twisted nematic) guest-host liquid crystal film has also been utilized in combination with a reflective polarizer to control the contrast of the scene [6]. Yet, simultaneous control of real-world-scene transmission and virtual-information brightness has not been demonstrated.

In this paper, we propose a HUD with electrically controlled contrast ratio, based on a cholesteric liquid crystal (CLC) [7–9] film, a geometry phase (GP) [10–14] liquid crystal (LC) lens and a circular polarizer. The CLC film here works as a reflective polarizer, which reflects one circular polarization with the same helical sense as the helix of the CLC, and simply transmits the orthogonal circular polarization. The GP LC lens [15–17] is a flat-plate diffractive device whose phase profile is generated by spatial variation of the LC directors instead of the optical path distance. Its diffraction efficiency is electrically tunable and is highly polarization-dependent. In our HUD system, by controlling voltage applied to the GP LC lens, the transmittance of the ambient light and the reflectance of the virtual light could be tuned simultaneously, so that the contrast ratio of the two can be controlled in an appropriate range. Therefore, the driver can always see virtual information and road conditions in different driving environments.

2. Principles of the CLC Film and the GP Lens

CLC is also called chiral nematic liquid crystal. Under the action of the chiral dopant, the nematic LC directors rotate along the normal direction of the plane, but in the same layer, all the LC directors are oriented in the same direction as shown in Figure 1a. The axial length over which the LC directors rotate by 180° is half pitch P/2. When light wavelength is within the reflection band of a CLC, the circularly polarized light with the same helical sense would be reflected, while the circularly polarized light with the opposite handedness would be transmitted. For both the reflected and transmitted light, polarization states are unchanged. For example, a right-handed CLC (RHCLC) exhibits a high reflection characteristic for right-handed circular polarization (RCP) and a high transmission characteristic for left-handed circular polarization (LCP) as shown in Figure 1b, and vice versa.

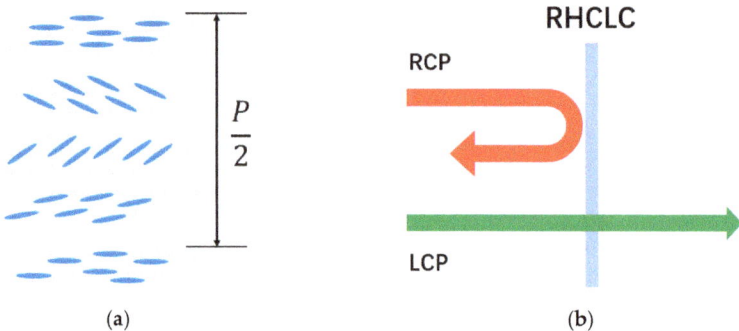

Figure 1. (a) Schematic diagram of the director configuration of a CLC; (b) Polarization-selective reflectivity of a CLC film.

The optic axis (LC directors) of a GP LC element change spatially, and that produces phase modulation for the incident light. The optic axis distribution of a GP lens in the substrate plane (x-y plane) is shown in Figure 2a. The azimuthal angle of the optic axis φ(x, y) changes continuously so that a parabolic phase profile is generated along the radial direction; in the direction perpendicular to the substrate (z-axis direction), the LC directors are uniformly aligned unless voltage is applied. When the incident light is a circular polarization, the output electric field vector after passing through a GP lens can be expressed as in [18]:

$$e^{i\delta_{in}} \begin{bmatrix} 1 \\ \pm i \end{bmatrix} \xrightarrow{GP} \sqrt{\eta} e^{i(\delta_{in} \pm 2\varphi)} \begin{bmatrix} 1 \\ \mp i \end{bmatrix} + \sqrt{\eta_0} e^{i\delta_{in}} \begin{bmatrix} 1 \\ \pm i \end{bmatrix}, \quad (1)$$

where $\delta_{in}(x, y)$ is the initial phase of the incident light. The output light consists of two components: a handedness inverted first-order circular polarization and a handedness unchanged zero-order circular polarization. The coefficient $\eta = \sin^2(\Gamma/2)$ is the first-order diffraction efficiency and $\eta_0 = \cos^2(\Gamma/2)$ is the zero-order diffraction efficiency, where Γ is the phase retardation. When half-wave retardation is satisfied ($\Gamma = \pi$), the first-order efficiency could theoretically reach 100% [13,19]. The phase shift term $\pm 2\phi$ in the first-order indicates that the sign of the additional phase is opposite for LCP and RCP. Therefore, a GP lens converges light of one circular polarization and diverges light of the other as illustrated in Figure 2b.

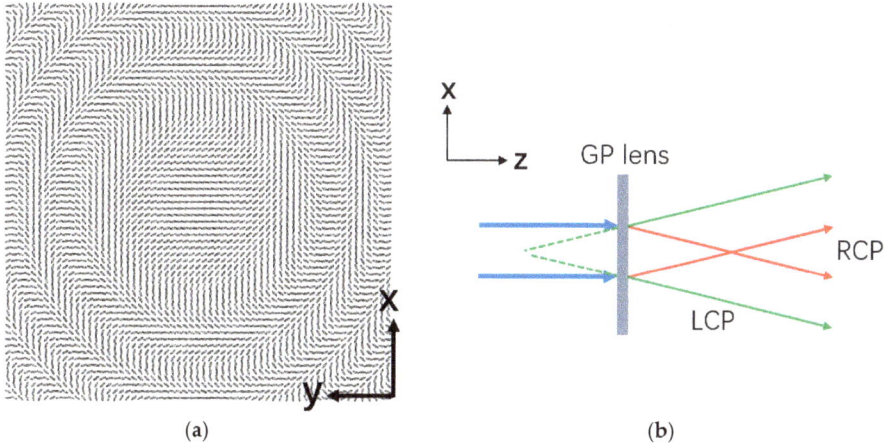

Figure 2. (a) Schematic distribution of LC directors in a GP lens; (b) Polarization dependency of a GP lens.

3. Working Principle of the HUD

The structural design of the HUD system we propose is shown in Figure 3. It consists of an image source, a broadband left-handed circular polarizer (Edmunds), a GP LC lens, and a RHCLC film. A circular polarizer could be considered as the optical combination of a polarizer and a quarter-wave plate, and in our structure, the polarizer side is facing the driver. The RHCLC is a reflective polarizer, which reflects most RCP and transmits most LCP.

For the virtual information light coming from the image source, it first encounters the left-handed circular polarizer and gets converted into the LCP. Next, the LCP light passes through the GP LC lens and gets diffracted. For this specific GP lens, its phase profile is designed in such a way that it exhibits a positive optical power for LCP light coming from the driver's side. So the first-order light is converged, and its handedness inversed, while the zero order remains left-handed as shown in Figure 4a. As they further proceed to the RHCLC film, the first order RCP is reflected towards the driver's side because of chirality match, while the zero order LCP is transmitted into the other side. Upon reflection, the first order light remains to be RCP thanks to the reflectivity characteristic of the CLC.

As the reflected RCP light encounters the GP LC lens for the second time, diffraction occurs again, which decomposes the RCP light into a LCP first-order component and a RCP zero-order component as shown in Figure 4b. Here, because both the light propagation direction and circular polarization handedness are reversed, the optical power of the lens remains positive, resulting in the first order light being even more converged. Eventually, the RCP zero-order is absorbed by the left-handed circular polarizer, while the converged LCP first-order manages to pass it through, and enters the driver's eye pupil.

For the virtual information light that reached the pupil, it passed through the positive GP lens twice. The effective optical power is approximately twice of that in a single passage. By appropriately choosing the image source location and the focal length of the GP LC lens, the virtual information can be magnified by the system. The diffraction efficiency through the double passage is approximately $\sin^4(\Gamma/2)$.

As for the ambient light, first of all, the RCP part of the light is reflected back by the RHCLC film, while the LCP part passes it through and encounters the elements that follow. Similarly, the GP lens diffracts the LCP ambient light into LCP zero-order light and RCP first-order. But the first-order light has diverged now as shown in Figure 4c, because the lens exhibits a negative focal length for LCP light coming from the external side. Finally, the left-handed circular polarizer filters out the RCP first-order, and only allows the LCP zero-order light to pass through.

From what has been discussed above, only the first-order of virtual information light and the zero-order of ambient light entering the human eye. By changing the applied voltage on the GP LC lens, the diffraction efficiencies for the first-order and zero-order will vary. Hence the brightness of the virtual information and ambient light could also be adjusted accordingly.

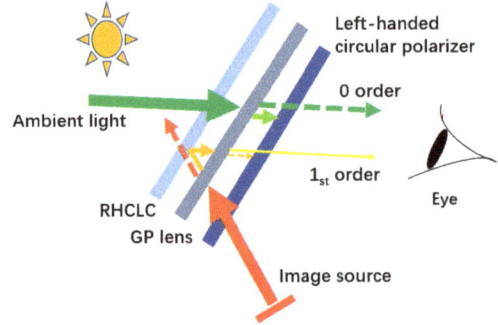

Figure 3. System design of the proposed augmented reality display.

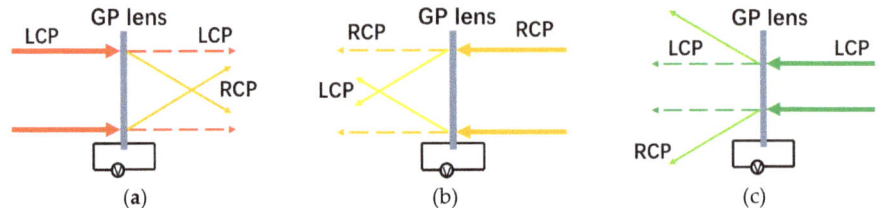

Figure 4. Virtual information light passing through the GP lens twice (**a**) the first time; (**b**) the second time; (**c**) Ambient light passing through the GP lens. The solid arrows represent first-order diffracted light, and the dashed arrows are zero-order light.

4. Experiment and Result

In our experiment, the RHCLC film was prepared by infiltrating an LC mixture containing 97.05 wt.% nematic liquid crystal with a large birefringence $\Delta n \sim 0.4$ (Changchun Institute of Optics, Changchun, Jilin, China) and 2.95 wt.% right-handed chiral dopant R5011 (HCCH, Nanjing, Jiangsu, China) into a 10 µm homogeneously aligned cell [20]. Using a high-resolution spectrometer (Ocean Optics, Shanghai, China), the transmittance spectrum of the RHCLC film was measured and shown in Figure 5. One can see that a flat reflection band within which high reflection occurs for RCP light ranges from 513 nm to 630 nm.

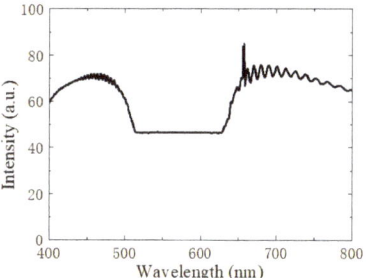

Figure 5. Transmittance spectrum of the RHCLC film.

The GP LC lens was fabricated using a non-interferometric single-exposure technique [21]. A 3 μm cell filled with an LC mixture consisted of 99 wt.% E7 (HCCH, Nanjing, Jiangsu, China) and 1 wt.% azo-dye methyl red (MR) (Sigma Aldrich, Shanghai, China). The size of the GP lens in our experiment was 1.5 cm × 0.9 cm. Figure 6a shows the micrograph of the GP lens under crossed polarizers. The total electrical switching time was 37.2 ms (rising time 27.8 ms and decay time 9.4 ms) as shown in Figure 6b. The rise time could be sped up further using the overdrive method [22]. The response time was fast enough to ensure quick contrast ratio adjustment when the ambient brightness changes.

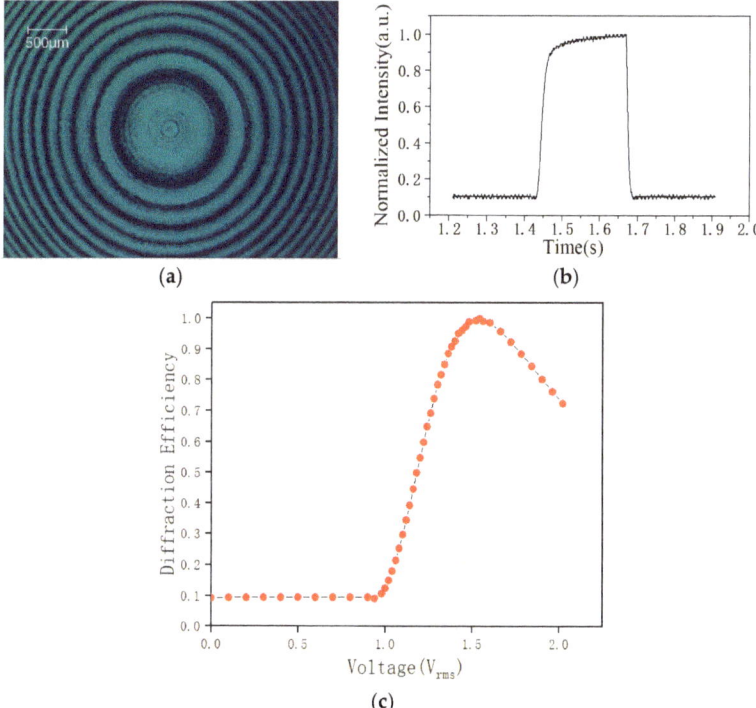

Figure 6. GP lens (a) optical micrograph under crossed polarizers; (b) response time; and (c) voltage-dependent first-order diffraction efficiency.

The voltage-dependent-diffraction efficiency (defined as the focused beam intensity divided by the total beam intensity behind the GP LC cell) for the wavelength 633 nm is shown in Figure 6c. Peak diffraction efficiency could be achieved at about 1.6 V_{rms}. Because the dye dopant MR has considerable absorption in blue and green, we purposely chose the color of virtual information light to be red.

Figure 7 shows the measured reflectance and transmittance of the system at different voltages. Surface reflections caused by the refractive index difference between the LC cell and the air can be reduced by gluing together each device except the image source with glycerin (n ~ 1.47). Since the virtual information is in red, the reflectance was measured using a 633 nm red laser. We also measured the transmittance using 633 nm, 543 nm, and 488 nm lasers to investigate the electro-optical properties of the GP LC lens for the ambient light. From the figure, one can see that for the same wavelength, the increase of reflectance is accompanied by the decrease of transmittance, because they correspond to the first-order and zero-order, respectively. However, due to the material dispersion and wavelength dependence of the GP effect itself, different wavelengths have different diffraction efficiencies. Since the human eye has the greatest visual sensitivity to green light, we take the curve of green light as the main reference. Over a range of voltage variations (approximately from 1.3 V_{rms} to 1.7 V_{rms}), the reflectance of the virtual information light increases while the transmittance of the ambient light decreases.

The transmittance curves for red, green, and blue light from the real world are mainly determined by the zero-order diffraction efficiency of the GP lens. Theoretically, when the applied voltage on the GP lens V is lower than the threshold voltage V_{th} [23], the zero-order diffraction efficiency (η_0) of the GP lens can be expressed as $\eta_0 = \cos^2(\pi\Delta n(\lambda)d/\lambda)$, where Δn is the birefringence of the LC, d is the cell gap, and λ is the wavelength. Here, d is ~ 3 µm and Δn of E7 for 633 nm is about 0.225, so the phase retardation term $\pi\Delta n(\lambda)d/\lambda$ is close to π, resulting in maximum efficiency η_0 for the red light at zero-voltage state. For green or blue light, $\pi\Delta n(\lambda)d/\lambda$ deviates from π, hence the transmittance is lower than that of red. In addition, the absorption of MR molecules for green and blue, and the transmittance spectrum of the CLC cell would also affect the absolute values of their transmittance.

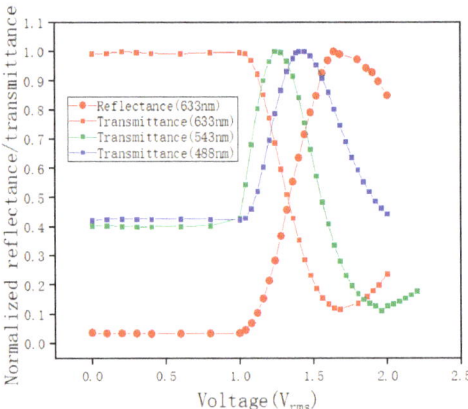

Figure 7. Voltage-dependent reflectance (632 nm laser) and transmittance (633 nm, 543 nm, and 488 nm lasers) of the system.

An LCD screen, with a red letter "S" in black background, was served as the light source. The distance between the LCD screen and the GP lens was 12cm. The GP lens and the CLC cell were stacked together with an index matching liquid, glycerin. A piece of paper with some black text was placed 23 cm from the CLC cell, on the other side of the system, as the real-world scene. A camera was placed at the pupil position to capture photos. Figure 8 shows the photos taken through the HUD by the camera. When the voltage applied on the GP lens increases gradually, the brightness of "S" increases and that of the background decreases. Because the virtual image is generated at a distance closer than the background, one can see from the figure that when "S" is in focus, the background is blurry as shown in the left part of the pictures, and when the background is in focus, the virtual "S" is blurry as shown in the right part. In real vehicles, when the eye focuses on the road, the road is absolutely clear. So it will not raise safety concerns.

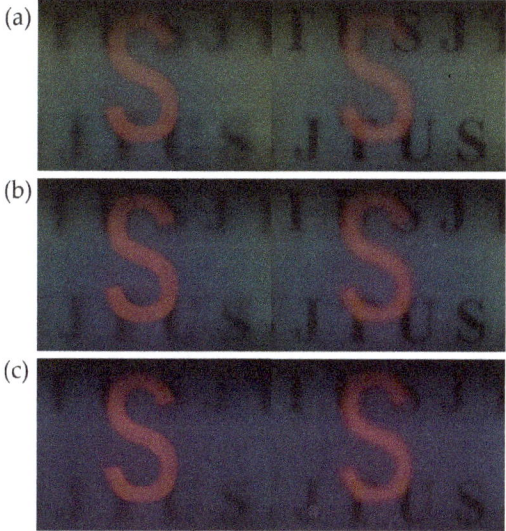

Figure 8. Photos focused on the virtual-information image and the background, respectively, at different voltages (**a**) 1.68 V$_{rms}$, (**b**) 1.80 V$_{rms}$, and (**c**) 1.96 V$_{rms}$.

5. Discussion

In our prototype, as the applied voltage increases, the color variation of the background accompanies the change of brightness. That is mainly because the diffraction efficiency of the GP lens is wavelength-dependent. To overcome this problem, one can use a dual-twist-structure GP lens in the system, which can achieve almost achromatic diffraction efficiency [24]. Here, as proof-of-concept verification, only a red virtual image was displayed by our proposed HUD. To realize full-color rendering, we plan to use SD1 [25], a photo-alignment material which has negligible absorption in the visible region, to replace MR in the future. Moreover, the reflection band of the CLC film should also be expanded to cover the entire visible range by using gradient pitch or multilayer methods [26]. Although the wavelength-dependent focal length (f) of GP lenses, which can be approximated by $f \propto 1/\lambda$ [23], would inevitably generate different magnifications for different colors, pre-calibration by the software in the image source could provide a simple yet effective solution to this. Considering the exotic features of the proposed HUD system, such as being able to adjust both ambient light and virtual information simultaneously, and the instant response to the applied voltage, it could greatly improve the virtual experience and driving safety. The application of such a display is not limited to HUDs in vehicles, but can also be extended to head-mounted displays for augmented reality.

6. Conclusions

We have proposed a HUD system for automobile applications, which can adjust the relative light intensity of ambient light and virtual light by changing the applied voltage. It provides a solution for HUDs to enable good visibility of both virtual information and real-world scenes under various driving conditions. The experimental results confirmed that dynamic control of brightness contrast could be realized within a voltage range of 1.68 V$_{rms}$ to 1.96 V$_{rms}$. In the future, the visual performance of the HUD could be further improved by expanding the reflection bandwidth of the CLC film and eliminating chromatic aberration of the GP lens.

Author Contributions: Conceptualization—Y.L. (Yan Li) and Y.L. (Yueda Liu); Formal analysis—Y.L. (Yueda Liu), Q.C. and S.L.; Funding acquisition—Y.L. (Yan Li) and Y.S.; Investigation—Y.L. (Yueda Liu); Project administration—Y.L. (Yan Li); Supervision—Y.L. (Yan Li); Validation—Y.L. (Yueda Liu); Writing – original draft—Y.L. (Yueda Liu); Writing, Review, and Editing—Y.L. (Yan Li).

Funding: National Natural Science Foundation of China (61727808); Shanghai Jiao Tong University (YG2016QN37); Key Lab of Advanced Optical Manufacturing Technologies of Jiangsu Province, and Key Lab of Modern Optical Technologies of Education Ministry of China, Soochow University (KJS1607).

Conflicts of Interest: The authors declare no conflict of interest.

References

1. How to Use BMW M Head-up Display. Available online: https://www.bmw-m.com/en/topics/magazine-article-pool/how-to-use-bmw-m-head-up-display.html (accessed on 16 June 2019).
2. Okumura, H.; Hotta, A.; Sasaki, T.; Horiuchi, K.; Okada, N. Wide field of view optical combiner for augmented reality head-up displays. In Proceedings of the 2018 IEEE International Conference on Consumer Electronics, Las Vegas, NV, USA, 12–14 January 2018; pp. 1–4.
3. Wei, S.; Fan, Z.; Zhu, Z.; Ma, D. Design of a head-up display based on freeform reflective systems for automotive applications. *Appl. Opt.* **2019**, *58*, 1675–1681. [CrossRef] [PubMed]
4. Zhan, T.; Lee, Y.-H.; Xiong, J.; Tan, G.; Yin, K.; Yang, J.; Liu, S.; Wu, S.-T. High-efficiency switchable optical elements for advanced head-up displays. *J. Soc. Inf. Disp.* **2019**, *27*, 223–231. [CrossRef]
5. Wirnsberger, G.; Scott, B.J.; Chmelka, B.F.; Stucky, G.D. Fast response photochromic mesostructures. *Adv. Mater.* **2000**, *12*, 1450–1454. [CrossRef]
6. Zhu, R.; Chen, H.; Kosa, T.; Coutino, P.; Tan, G.; Wu, S.T. High-ambient-contrast augmented reality with a tunable transmittance liquid crystal film and a functional reflective polarizer. *J. Soc. Inf. Disp.* **2016**, *24*, 229–233. [CrossRef]
7. Wu, S.-T.; Yang, D.-K. *Reflective Liquid Crystal Displays*; Wiley: New York, NY, USA, 2001.
8. Mitov, M.; Dessaud, N. Going beyond the reflectance limit of cholesteric liquid crystals. *Nat. Mater.* **2006**, *5*, 361–364. [CrossRef] [PubMed]
9. Yang, D.-K.; Wu, S.-T. *Fundamentals of Liquid Crystal Devices*; Wiley: New York, NY, USA, 2007.
10. Anandan, J. The geometric phase. *Nature* **1992**, *360*, 307–313. [CrossRef]
11. Pancharatnam, S. Generalized theory of interference, and its applications. *Proc. Indian Acad. Sci. A* **1956**, *44*, 398–417. [CrossRef]
12. Berry, M.V. Quantal phase factors accompanying adiabatic changes. *Proc. R. Soc. Lond.* **1987**, *A392*, 45–57.
13. Marrucci, L.; Manzo, C.; Paparo, D. Pancharatnam-Berry phase optical elements for wavefront shaping in the visible domain: Switchable helical modes generation. *Appl. Phys. Lett.* **2006**, *88*, 221102. [CrossRef]
14. Zheng, G.; Muhlenbernd, H.; Kenney, M.; Li, G.; Zentgraf, T.; Zhang, S. Metasurface holograms reaching 80% efficiency. *Nat. Nanotechnol.* **2015**, *10*, 308–312. [CrossRef] [PubMed]
15. Zhan, T.; Lee, Y.-H.; Wu, S.-T. High-resolution additive light field near-eye display by switchable Pancharatnam–Berry phase lenses. *Opt. Express* **2018**, *26*, 4863–4872. [CrossRef] [PubMed]
16. Tan, G.; Zhan, T.; Lee, Y.-H.; Xiong, J.; Wu, S.-T. Polarization-multiplexed multiplane display. *Opt. Lett.* **2018**, *43*, 5651–5654. [CrossRef] [PubMed]
17. Lee, Y.-H.; Tan, G.; Yin, K.; Zhan, T.; Wu, S.-T. Compact see-through near-eye display with depth adaption. *J. Soc. Inf. Disp.* **2018**, *26*, 64–70. [CrossRef]
18. Kim, J.; Li, Y.; Miskiewicz, M.N.; Oh, C.; Kudenov, M.W.; Escuti, M.J. Fabrication of ideal geometric-phase holograms with arbitrary wavefronts. *Optica* **2015**, *11*, 958–964. [CrossRef]
19. Nikolova, L.; Ramanujam, P.S. *Polarization Holography*; Cambridge University Press: Cambridge, UK, 2009.
20. Chen, Q.; Peng, Z.; Li, Y.; Liu, S.; Zhou, P.; Gu, J.; Lu, J.; Yao, L.; Wang, M.; Su, Y. Multi-plane augmented reality display based on cholesteric liquid crystal reflective films. *Opt. Express* **2019**, *27*, 12039–12047. [CrossRef] [PubMed]
21. Li, Y.; Liu, Y.; Li, S.; Zhou, P.; Zhan, T.; Chen, Q.; Su, Y.; Wu, S.-T. Single-exposure fabrication of tunable Pancharatnam-Berry devices using a dye-doped liquid crystal. *Opt. Express* **2019**, *27*, 9054–9060. [CrossRef] [PubMed]
22. Wu, S.T. Nematic liquid crystal modulator with response time less than 100 µs at room temperature. *Appl. Phys. Lett.* **1990**, *57*, 986–988. [CrossRef]

23. Lee, Y.-H.; Tan, G.; Zhan, T.; Weng, Y.; Liu, G.; Gou, F.; Peng, F.; Tabiryan, N.V.; Gauza, S.; Wu, S.-T. Recent progress in Pancharatnam–Berry phase optical elements and the applications for virtual/augmented realities. *Opt. Data Process. Storage* **2017**, *3*, 79–88. [CrossRef]
24. Oh, C.; Escuti, M.J. Achromatic diffraction from polarization gratings with high efficiency. *Opt. Lett.* **2008**, *33*, 2287–2289. [CrossRef] [PubMed]
25. Chigrinov, V.G.; Kwok, H.; Yip, W.C.; Kozenkov, V.M.; Prudnikova, E.K.; Tang, B.Z.; Salhi, F. New photo-aligning and photo-patterning technology: Superthin internal polarizers, retarders, and aligning layers. In Proceedings of the International Symposium on Optical Science and Technology, San Diego, CA, USA, 11 December 2001; pp. 117–132.
26. Mitov, M. Cholesteric Liquid Crystals with a Broad Light Reflection Band. *Adv. Mater.* **2012**, *24*, 6260. [CrossRef] [PubMed]

© 2019 by the authors. Licensee MDPI, Basel, Switzerland. This article is an open access article distributed under the terms and conditions of the Creative Commons Attribution (CC BY) license (http://creativecommons.org/licenses/by/4.0/).

Communication

Electro-Optical Properties of a Polymer Dispersed and Stabilized Cholesteric Liquid Crystals System Constructed by a Stepwise UV-Initiated Radical/Cationic Polymerization

Chen-Yue Li [1,†], Xiao Wang [2,†], Xiao Liang [3], Jian Sun [3], Chun-Xin Li [3], Shuai-Feng Zhang [1], Lan-Ying Zhang [3], Hai-Quan Zhang [2,*] and Huai Yang [3,*]

1. Department of Materials Physics and Chemistry, School of Materials Science and Engineering, University of Science and Technology Beijing, Beijing 100083, China; li.rk@163.com (C.-Y.L.); shuaifengustb@163.com (S.-F.Z.)
2. State Key Laboratory of Metastable Materials Science and Technology, Yanshan University, Qinhuangdao 066004, China; wxdxasdzx@163.com
3. Department of Materials Science and Engineering, College of Engineering, Peking University, Beijing 100871, China; hitliangxiao@163.com (X.L.); sun6jian10@gmail.com (J.S.); xin41264012@163.com (C.-X.L.); zhanglanying@pku.edu.cn (L.-Y.Z.)
* Correspondence: hqzhang@ysu.edu.cn (H.-Q.Z.); yanghuai@pku.edu.cn (H.Y.); Tel.: +86-010-62766919 (H.Y.)
† These authors contributed equally to this work.

Received: 20 March 2019; Accepted: 22 May 2019; Published: 29 May 2019

Abstract: Polymer-dispersed liquid crystal (PDLC) and polymer-stabilized liquid crystal (PSLC) are two typical liquid crystal (LC)/polymer composites. PDLCs are usually prepared by dispersing LC droplets in a polymer matrix, while PSLC is a system in which the alignment of LC molecules is stabilized by interactions between the polymer network and the LC molecules. In this study, a new material system is promoted to construct a coexistence system of PDLC and PSLC, namely PD&SChLC. In this new material system, a liquid-crystalline vinyl-ether monomer (LVM) was introduced to a mixture containing cholesteric liquid crystal (ChLC) and isotropic acrylate monomer (IAM). Based on the different reaction rates between LVM and IAM, the PD&SChLC architecture was built using a stepwise UV-initiated polymerization. During the preparation of the PDS&ChLC films, first, the mixture was irradiated with UV light for a short period of time to induce the free radical polymerization of IAMs, forming a phase-separated microstructure, PDLC. Subsequently, an electric filed was applied to the sample for long enough to induce the cationic polymerization of LVMs, forming the homeotropically-aligned polymer fibers within the ChLC domains, which are similar to those in a PSLC. Based on this stepwise UV-initiated radical/cationic polymerization, a PD&SChLC film with the advantages of a relatively low driving voltage, a fast response time, and a large-area processability is successful prepared. The film can be widely used in flexible displays, smart windows, and other optical devices.

Keywords: cholesteric liquid crystals; polymer; radical polymerization; cationic polymerization; electro-optical property; microstructure

---

1. Introduction

Polymer-dispersed cholesteric liquid crystal (PDLC) and polymer-stabilized cholesteric liquid crystal (PSChLC) systems are two important classes of liquid crystal (LC)/polymer composite materials [1–9]. The PDLC films exhibit a micro phase separation structure when LC droplets are uniformly dispersed into the porous polymer matrix. In normal mode PDLC films scatter light, while on application of an external electric field they can be switched into a transparent state [10–14]. Owing to their advantages of facile formation, good film formability, and stable optical properties, PDLCs are used in a wide variety of applications, such as curtain free windows, displays, micro lenses, and light shutters, etc. [15]. However, the polymer network in the PDLCs lack directional orientation due to the weaker molecular interactions between the ChLCs and the polymer. The reorientation of LCs in the polymer network requires a high driving voltage.

On the other hand, an orientated liquid crystalline polymer network of LCs in a PSChLC system can facilitate cholesteric liquid crystal molecule re-orientation from a focal conic texture to homeotropic alignments upon the application of an electric field [16,17]. Thus, PSChLC requires a low driving voltage and has a prompt response rate. However, the low polymer content in PSChLC is usually less than 10%, causing poor mechanical strength [18–20]. Recently, a coexistent system combining the advantages of both polymer-dispersed and polymer-stabilized liquid crystals (PD&SLCs) has been developed by our group. This hybrid system is garnering substantial popularity in both academia and the industry due to its unique microstructure [21]. The produced electrically switchable PD&SChLC film not only shows a more than 50% decrease in driving voltage compared with conventional PDLC films but also possess high mechanical strength.

Generally, a two-step polymerization strategy is pursued to accomplish a PD&SChLC system. As reported in the literature, in the first step a porous polymer network is obtained by UV irradiation of isotropic acrylate monomers (IAMs). Secondly, an electric field is applied simultaneously with UV-light exposure, the liquid crystal acrylate monomers (LAMs) within the ChLC domains being crosslinked to form homeotropically aligned polymer fibers in the porous matrix [22–25]. During the first stage some of the LAMs are consumed by radical polymerization under UV light. To leave sufficient LAMs for forming the oriented polymer fibers in the second UV polymerization step, the curing time of UV irradiation in the first stage should be strictly controlled, which is not beneficial for the polymer morphology control or further optimization of the electro-optical properties of the PD&SChLC film [22–25]. To date, liquid-crystalline vinyl-ether monomers (LVMs) have not been investigated in constructing a PD&SChLC system.

In this study, we propose a new materials system containing LVMs, ChLCs, and IAMs, and a PD&SChLC architecture is built stepwise by first consuming the IAMs via free radical polymerization, followed by crosslinking the LVMs within the ChLC domains via cationic polymerization. Due to the different rates between the radical reaction and the cationic one, the two polymerization steps are completely separated, and a coexist structure with homeotropically aligned polymer fibers inside the porous polymer matrix is formed.

2. Materials and Methods

2.1. Materials

Figure 1 shows the chemical structures and physical parameters of the materials used in this study. Unless otherwise specified, the reagents were purchased from Sigma-Aldrich (St. Louis, USA). Transparent and conductive polyester substrates were gifted from Dalian Jingcai Smart Film Technology Co., Ltd. (Dalian, China). IAMs composed of lauryl methacrylate (LMA), polyethylene glycol diacrylate (PEGDA 600), and hydroxypropyl methacrylate (HPMA) were purchased from Tokyo Chemical Industry Co., Ltd. (Tokyo, Japan). LVMs and C4V were purchased from Beijing Green Innov-Tech Co., Ltd. (Beijing, China). Photo-initiators (PI), UVI-6976 and Irgacure 651 were bought, respectively, from Synasia (Suzhou) Co., Ltd. and TCI Co., Ltd. (Suzhou, China). Nematic LCs and SLC-1717 were purchased from Shijiazhuang Yongsheng Huatsing Liquid Crystal Co., Ltd. (Shijiazhuang, China). The chiral dopant, S811, was obtained from Beijing Gerui Technology Co., Ltd. (Beijing, China). Glass spacers with a diameter of 20 μm were obtained from Sekisui Chemical Co., Ltd. (Osaka, Japan) to control the thickness of the film.

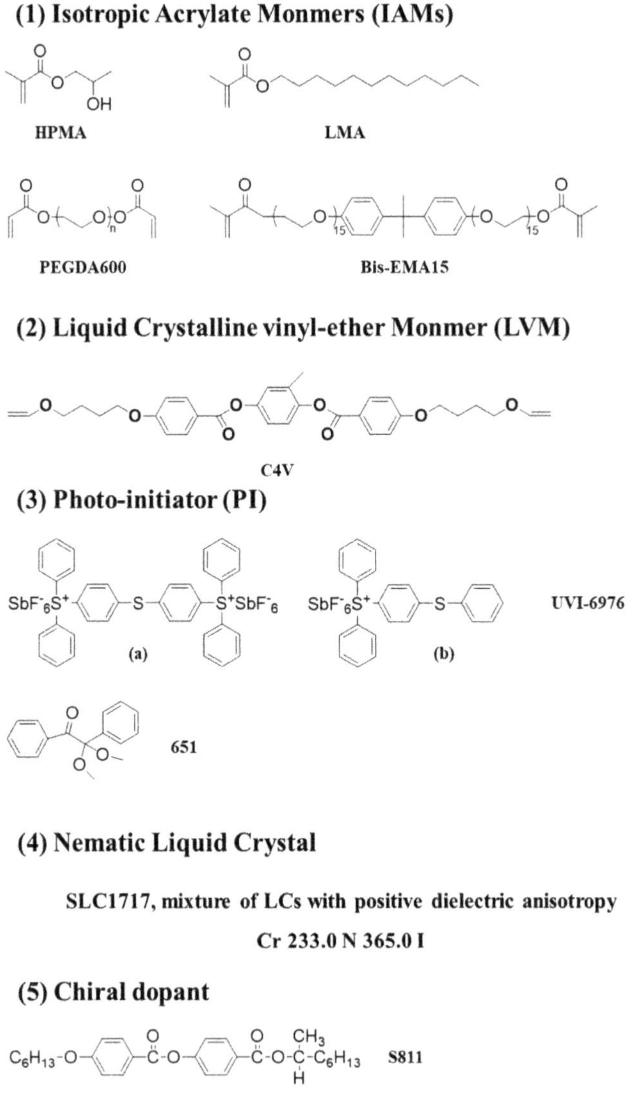

(1) Isotropic Acrylate Monmers (IAMs)

HPMA

LMA

PEGDA600

Bis-EMA15

(2) Liquid Crystalline vinyl-ether Monmer (LVM)

C4V

(3) Photo-initiator (PI)

SbF$_6^-$S$^+$ ——S—— S$^+$SbF$_6^-$ SbF$_6^-$S$^+$ ——S——   UVI-6976

(a)   (b)

651

(4) Nematic Liquid Crystal

SLC1717, mixture of LCs with positive dielectric anisotropy

Cr 233.0 N 365.0 I

(5) Chiral dopant

C$_6$H$_{13}$-O-...-C-O-...-C-O-C-C$_6$H$_{13}$ S811

Figure 1. Chemical structures and physical properties of the materials used.

The composition of the IAMs is shown in Table S1. The ChLC comprises SLC177 and S811 and the corresponding composition is listed in Table S2. Meanwhile, the polarizing optical microscope (POM) figures of the ChLC are shown in Figure S1. The PI used in this study are UVI-6976 and Irgacure 651; theor composition is listed in Table S3. All these materials were used without further purification.

2.2. Preparations of the Samples

Films A, B, C, and D, whose compositions are listed in Table 1, were prepared according to the procedures schematically illustrated in Figure 2. Firstly, a mixture of LVMs, ChLC, IAMs, photo-initiators, and glass spacers was stirred evenly until it became transparent. Then, the mixture was filled into two transparent and conductive polyester substrates via capillary action. Following this, the samples were cured under a UV light (365 nm, 3.0 mW/cm^2) for about 80 s to form the PDLC-like porous polymer structure. Next, a 100V, 1000Hz square wave electric field was applied across the samples to switch them into a transparent state. Simultaneously, a second-step UV light (365 nm) curing with the aforementioned intensity was employed for 10 min to fully consume the LVMs within the ChLC domains to form homeotropically aligned polymer fibers in the porous polymer structure. It is worth mentioning that all the films were prepared at room temperature.

Table 1. Compositions of the films with different LVM contents.

Sample	ChLC (wt%)	LVMs (wt%)	IAMs (wt%)	PI (wt%)
A	64.5	0	35.0	0.5
B	64.0	0.5	35.0	0.5
C	63.5	1.0	35.0	0.5
D	63.0	1.5	35.0	0.5

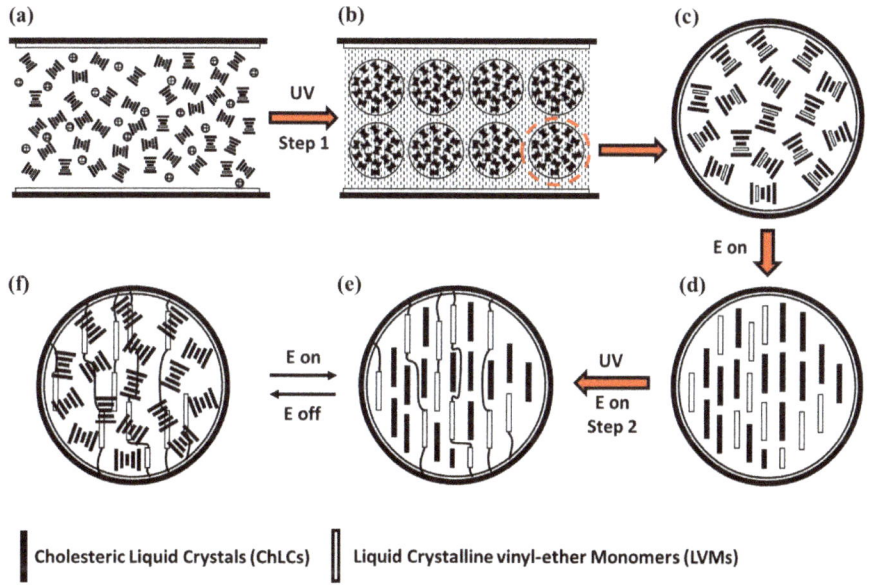

Figure 2. Schematic illustration for the preparation of the PD&SChLC film.

2.3. Measurements

The E-O properties of the samples were measured using a liquid crystal device parameter tester (LCT-5016C, Changchun Liancheng Instrument Co. Ltd., Changchun, China). Each film was applied with a square wave-modulated electric field with a frequency of 100 Hz during the measurement. The transmittance of conductive polyester substrates was normalized as 100%.

The morphologies of the polymer structure were observed via scanning electron microscopy (SEM, HITACHI S-4800, Tokyo, Japan). Cross-sectional SEM images of the samples were obtained by firstly breaking off the samples in liquid nitrogen and then eliminating the LC molecules by immersing the samples in a cyclohexane solvent for about seven days. After that, the samples were dried in an oven at 60 °C for about 3 h and coated with a thin layer of gold before SEM observation.

3. Results and Discussion

Figure 2 shows a schematic illustration of tailoring the PD&SChLC architecture in the film. Firs, a well-stirred mixture of ChLC, IAMs, and LVMs is sandwiched between two transparent and conductive polyester substrates (Figure 2a). The film shows a transparent state which is attributed to the isotropic nature of the mixture. After the film is irradiated, the porous polymer structure built by the IAMs is roughly formed, as shown in Figure 2b, in the first 80 s. The SEM photograph of the porous polymer morphology is also shown in Figure S2. Meanwhile, a preliminary phase separation between the cross-linked polymer and ChLCs (including LVMs) also occurs with the consumption of the IAMs. In this phase separation process, the optical appearance of the film turns from a transparent into a light scattering state due to the focal conic textures of the ChLCs, as shown in Figure 2c. It is worth mentioning that the polymerization rate of the acrylic monomer (which belongs to the radical reaction) is much faster than that of the vinyl ether monomers. Thus, the LVMs remain completely unpolymerized after this first step of UV irradiation [26–28].

Subsequently, an electric field is applied to homeotropically orient the ChLCs and the LVMs, and, accordingly, the film turns back to a transparent state, as shown in Figure 2d. Simultaneously, a second stage of UV polymerization is carried out using a high light intensity to consume the LVMs via cationic polymerization in order to form homeotropically oriented polymer fibers within the ChLC droplets, as shown in Figure 2e. During the second step the films are kept transparent via an applied electric field to ensure the homeotropic alignment of the LVMs. After the removal of the electric field, the film turns back to a light scattering state because the ChLC molecules return from the homeotropic alignments to the focal conic textures, as shown in Figure 2f.

To verify the coexistent PD&SChLC structure, SEM photographs of the polymer microstructures of the as-made films, from horizontal and cross-sectional perspectives, were taken. As shown in Figure 3a,c, the microstructures of the PDChLC film have a polymer matrix with a uniformed porous structure, with the size of each pore being about 3–5 μm, which is in agreement with typical PDLC materials. By comparing Figure 3a,b, we can see that with the introduction of LVMs, the homeotropically oriented polymer fibers can be successfully formed in the new materials system. Figure 3b shows the enlarged view of the coexistent porous polymer structure and the oriented polymer fibers. As shown in Figure 3d, the morphologies of the polymer configuration from the horizontal perspective are also distinct from those of the PD&SChLC film, compared with Figure 3c. From the above characterization, we can confirm that the coexistent PD&SChLC structure can be successfully achieved by a promoted stepwise UV-initiated radical/cationic polymerization strategy.

According to the above investigation, we deduced that the homeotropically-oriented polymer fibers formed by LVMs may bring new electro-optical properties to the as-made PD&SChLC film. Herein, a series of the PD&SChLC films were prepared by varying the LVMs contents, with the compositions of the films listed in Table 1. The results of the electro-optic properties of these films are summarized in Figure 4a,b. As the LVM content increases from 0 wt% to 1.5 wt%, the threshold voltage (V_{th}) and the saturation voltage (V_{sat}) of the films, which are defined as the voltage required when the film transmittance reaches 10% and 90% of maximum transmittance, respectively, are greatly reduced. Specifically, the V_{th} and V_{sat} of the films decreases from 43.3 V/88.7 V to 32.2 V/58.8 V, 25.5 V/44.6 V, and 21.2 V/35.5 V, as shown in Figure 4b. This electro-optical improvement can be attributed to the homeotropically oriented polymer fibers, which may reduce the anchoring force of the porous network while rotating the ChLC molecules through the applied electric field. In addition, obviously, the anchoring force can be further reduced with more polymer fibers formed via increasing the LVM content. If the LVM content is increased up to 2.0 wt%, the transmittance of the film under normal conditions will be greatly increased, which results from the stabilization effects of the high densities of the homeotropically-oriented polymer fibers, which reduce the contrast ratios (CR) of the films (Figure S3). Hence, the electro-optic properties of the films reached their best when the content of the LVMs was 1.5 wt% in our experiments.

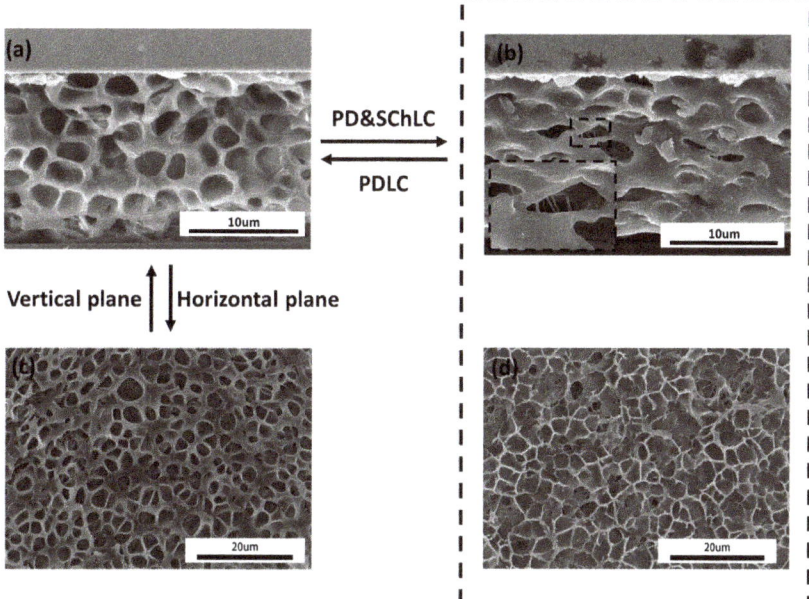

Figure 3. Scanning electron microscopy (SEM) photographs of the polymer microstructures for the as-made films from horizontal and cross-sectional perspectives: (**a**,**c**) are from the PDLC film And (**b**,**d**) are from the PD&SChLC film.

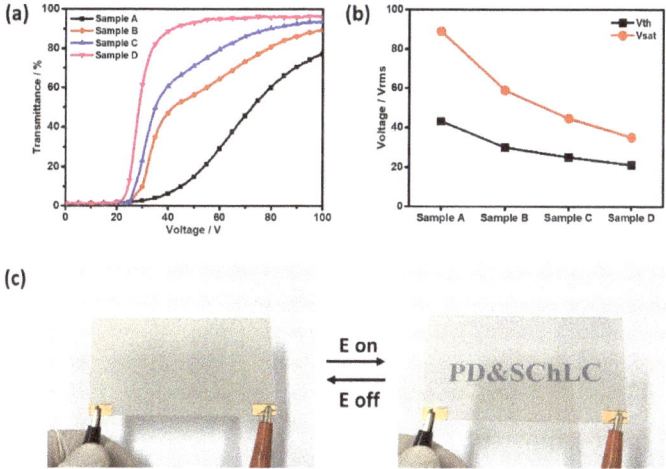

Figure 4. (**a**) Applied voltage dependence of transmittance of the as-made films containing A 0 wt%, B 0.5 wt%, C 1.0 wt%, and D 1.5 wt% LVMs, respectively. (**b**) The threshold voltage (Vth) and saturation voltage (Vsat) of the films. (**c**) Digital photographs showing the arbitrary transformation of Sample D from an intense light scattering state (left) to a highly transparent state (right) under electric field control.

Moreover, compared to the previously promoted PD&SChLC system, which had 3.0 wt% liquid-crystalline acrylate monomer content [21–25], having 1.5% LVM content in the new PD&SChLC system is the result of the two completely separated polymerizations steps. This is what we envisioned by introducing two means of aggregation, these being free radical and cationic.

Ultimately, the prepared films show a strong light scattering state under normal conditions, as shown in Figure 4c, and can achieve a fast transition to a highly transparent state under an electric field lower than 40 V. In addition, the response time for the electro-optical transition between the two states is fast, and the total time of the roundtrip is less than 30 ms, which is proved in Figure S4.

4. Conclusions

In conclusion, in this work a novel strategy for the preparation of PD&SChLC films by a stepwise UV-initiated free radical/cationic polymerization is proposed and validated. The new material system, containing LVMs, ChLCs and IAMs, is promoted to prepare a PD&SChLC film which has the advantages of low-driving voltage, easy preparation, good mechanical properties, and a fast response rate. Due to the reaction rate of the cationic polymerization being far slower than the radical polymerization, the two UV polymerization processes in the new system are able to be truly separated. Based on this, moving from 3.0 wt% to 1.5 wt%, the content of the LVMs can be decreased exponentially compared to the IAMs in the previous material system. Meanwhile, the homeotropically oriented polymer fibers constructed by the LVMs can effectively decrease the Vth and Vsat of the films from 43.3 V/88.7 V to 21.2 V/35.5 V. With these advantages, the new PD&SChLC system can be widely applicated in flexible displays, smart windows, and other optical devices.

Supplementary Materials: The following are available online at http://www.mdpi.com/2073-4352/9/6/282/s1, Table S1. Weight ratio of the ChLC. Table S2. Molar/weight ratio of the IAMs we used in this work. Table S3. Weight ratio of the photo-initiator. Figure S1. POM textures of SLC 1717 and ChLC (97 wt% SLC1717 + 3 wt% S811). Figure S2. SEM photograph of the porous network in 80s. Figure S3. The contrast ratio of the films with different LVM contents. Figure S4. Response time of switching state conversion of the film.

Author Contributions: C.-Y.L. and X.W. contributed equally to this work. Conceptualization, C.-Y.L. and X.W.; methodology, X.L.; validation, C.-Y.L., X.W., and J.S.; formal analysis, C.-X.L. and S.-F.Z.; investigation, L.-Y.Z.; writing—original draft preparation, X.W.; writing—review and editing, C.-Y.L., X.W., and X.L.; project administration, H.-Q.Z. and H.Y.; funding acquisition, H.Y.

Funding: This research was funded by the National Key R&D Program of China (grant no. 2018YFB0703704), the National Natural Science Foundation of China (NSFC) (grant nos. 51720105002, 51561135014, and 51573003), and the Joint Fund of the Ministry of Education for Equipment Pre-Research (grant nos. 6141A020222 and 6141A020332).

Conflicts of Interest: The authors declare no conflict of interest.

References

1. Higgins, D.A. Probing the mesoscopic chemical and physical properties of polymer-dispersed liquid crystals. *Adv. Mater.* **2000**, *12*, 251–264. [CrossRef]
2. Kumano, N.; Seki, T.; Ishii, M.; Nakamura, H.; Umemura, T.; Takeoka, Y. Multicolor Polymer-Dispersed Liquid Crystal. *Adv. Mater.* **2011**, *23*, 884–888. [PubMed]
3. Kim, M.; Park, K.J.; Seok, S.; Ok, J.M.; Jung, H.-T.; Choe, J.; Kim, D.H. Fabrication of microcapsules for dye-doped polymer-dispersed liquid crystal-based smart windows. *ACS Appl. Mater. Interfaces* **2015**, *7*, 17904–17909. [CrossRef]
4. Kikuchi, H.; Yokota, M.; Hisakado, Y.; Yang, H.; Kajiyama, T. Polymer-stabilized liquid crystal blue phases. *Nat. Mater.* **2002**, *1*, 64. [CrossRef] [PubMed]
5. Broer, D.; Lub, J.; Mol, G. Wide-band reflective polarizers from cholesteric polymer networks with a pitch gradient. *Nature* **1995**, *378*, 467. [CrossRef]
6. Dierking, I. Polymer network—Stabilized liquid crystals. *Adv. Mater.* **2000**, *12*, 167–181. [CrossRef]
7. Vaz, N.A.; Smith, G.W.; Montgomery, G.P., Jr. A Light Control Film Composed of Liquid Crystal Droplets Dispersed in a UV-Curable Polymer. *Mol. Cryst. Liq. Cryst.* **1987**, *146*, 1–15. [CrossRef]
8. Doane, J.W.; Golemme, A.; West, J.L.; Whitehead, J.B.; Wu, B.G. Polymer Dispersed Liquid Crystals for Display Application. *J. Funct. Polym.* **1990**, *165*, 511–532. [CrossRef]
9. Le, Z.; Ma, H.; Cheng, H.; Wei, H.; Zhang, S.; Zhang, L.; Yang, H. A novel light diffuser based on the combined morphology of polymer networks and polymer balls in a polymer dispersed liquid crystals film. *RSC Adv.* **2018**, *8*, 21690–21698.

10. Nastał, E.; Żurańska, E.; Mucha, M. Effect of curing progress on the electrooptical and switching properties of PDLC system. *J. Appl. Polym. Sci.* **1999**, *71*, 455–463. [CrossRef]
11. Vaz, N.A.; Montgomery, G.P., Jr. Refractive indices of polymer-dispersed liquid-crystal film materials: Epoxy-based systems. *J. Appl. Phys.* **1987**, *62*, 3161–3172. [CrossRef]
12. Higgins, D.A.; Hall, J.E.; Xie, A. Optical microscopy studies of dynamics within individual polymer-dispersed liquid crystal droplets. *Acc. Chem. Res.* **2005**, *38*, 137–145. [CrossRef] [PubMed]
13. Meng, Q.; Cao, H.; Kashima, M.; Liu, H.; Yang, H. Effects of the structures of epoxy monomers on the electro-optical properties of heat-cured polymer-dispersed liquid crystal films. *Liq. Cryst.* **2010**, *37*, 189–193. [CrossRef]
14. Filpo, G.D.; Formoso, P.; Mashin, A.I.; Nezhdanov, A.; Mochalov, L.; Nicoletta, F.P. A new reverse mode light shutter from silica-dispersed liquid crystals. *Liq. Cryst.* **2018**, *45*, 721–727. [CrossRef]
15. Zhang, C.; Wang, D.; Cao, H.; Song, P.; Yang, C.; Yang, H.; Hu, G.H. Preparation and electro-optical properties of polymer dispersed liquid crystal films with relatively low liquid crystal content. *Polym. Adv. Technol.* **2013**, *24*, 453–459. [CrossRef]
16. Dierking, I.; Kosbar, L.; Afzali-Ardakani, A.; Lowe, A.; Held, G. Two-stage switching behavior of polymer stabilized cholesteric textures. *J. Appl. Phys.* **1997**, *81*, 3007–3014. [CrossRef]
17. Yang, D.K.; Chien, L.C.; Doane, J. Cholesteric liquid crystal/polymer dispersion for haze-free light shutters. *Appl. Phys. Lett.* **1992**, *60*, 3102–3104. [CrossRef]
18. Rajaram, C.V.; Hudson, S.; Chien, L. Morphology of polymer-stabilized liquid crystals. *Chem. Mater.* **1995**, *7*, 2300–2308. [CrossRef]
19. Broer, D.J.; Boven, J.; Mol, G.N.; Challa, G. In-Situ photopolymerization of oriented liquid-crystalline acrylates, 3. Oriented polymer networks from a mesogenic diacrylate. *Die Makromol. Chem.* **1989**, *190*, 2255–2268. [CrossRef]
20. Muzic, D.; Rajaram, C.; Chien, L.; Hudson, S. Morphology of polymer networks polymerized in highly ordered liquid crystalline phases. *Polym. Adv. Technol.* **1996**, *7*, 737–742. [CrossRef]
21. Guo, S.-M.; Liang, X.; Zhang, C.-H.; Chen, M.; Shen, C.; Zhang, L.-Y.; Yuan, X.; He, B.-F.; Yang, H. Preparation of a thermally light-transmittance-controllable film from a coexistent system of polymer-dispersed and polymer-stabilized liquid crystals. *ACS Appl. Mater. Interfaces* **2017**, *9*, 2942–2947. [CrossRef] [PubMed]
22. Liang, X.; Guo, S.; Chen, M.; Li, C.; Wang, Q.; Zou, C.; Zhang, C.; Zhang, L.; Guo, S.; Yang, H. A temperature and electric field-responsive flexible smart film with full broadband optical modulation. *Mater. Horiz.* **2017**, *4*, 878–884. [CrossRef]
23. Liang, X.; Guo, C.; Chen, M.; Guo, S.; Zhang, L.; Li, F.; Guo, S.; Yang, H. A roll-to-roll process for multi-responsive soft-matter composite films containing Cs_xWO_3 nanorods for energy-efficient smart window applications. *Nanoscale Horiz.* **2017**, *2*, 319–325. [CrossRef]
24. Liang, X.; Chen, M.; Guo, S.; Zhang, L.; Li, F.; Yang, H. Dual-band modulation of visible and near-infrared light transmittance in an all-solution-processed hybrid micro–nano composite film. *ACS Appl. Mater. Interfaces* **2017**, *9*, 40810–40819. [CrossRef]
25. Liang, X.; Chen, M.; Guo, S.; Wang, X.; Zhang, S.; Zhang, L.; Yang, H. Programmable electro-optical performances in a dual-frequency liquid crystals/polymer composite system. *Polymer* **2018**, *149*, 164–168. [CrossRef]
26. Voytekunas, V.Y.; Ng, F.L.; Abadie, M.J. Kinetics study of the UV-initiated cationic polymerization of cycloaliphatic diepoxide resins. *Eur. Polym. J.* **2008**, *44*, 3640–3649. [CrossRef]
27. Decker, C.; Viet, T.N.T.; Decker, D.; Weber-Koehl, E. UV-radiation curing of acrylate/epoxide systems. *Polymer* **2001**, *42*, 5531–5541. [CrossRef]
28. Crivello, J.V. The discovery and development of onium salt cationic photoinitiators. *J. Polym. Sci. Part. A Polym. Chem.* **1999**, *37*, 4241–4254. [CrossRef]

© 2019 by the authors. Licensee MDPI, Basel, Switzerland. This article is an open access article distributed under the terms and conditions of the Creative Commons Attribution (CC BY) license (http://creativecommons.org/licenses/by/4.0/).

Article

Nematic and Cholesteric Liquid Crystal Structures in Cells with Tangential-Conical Boundary Conditions

Mikhail N. Krakhalev [1,2,*], Rashid G. Bikbaev [1,2], Vitaly S. Sutormin [1], Ivan V. Timofeev [1,2] and Victor Ya. Zyryanov [1]

[1] Kirensky Institute of Physics, Federal Research Center KSC SB RAS, Krasnoyarsk 660036, Russia; bikbaev@iph.krasn.ru (R.G.B.); sutormin@iph.krasn.ru (V.S.S.); tiv@iph.krasn.ru (I.V.T.); zyr@iph.krasn.ru (V.Y.Z.)
[2] Siberian Federal University, Krasnoyarsk 660041, Russia
* Correspondence: kmn@iph.krasn.ru; Tel.: +7-391-249-4510

Received: 27 March 2019; Accepted: 9 May 2019; Published: 14 May 2019

Abstract: Orientational structures formed in nematic and cholesteric layers with tangential-conical boundary conditions have been investigated. LC cells with one substrate specifying the conical surface anchoring and another substrate specifying the tangential one have been considered. The director configurations and topological defects have been identified analyzing the texture patterns obtained by polarizing microscope in comparison with the structures and optical textures calculated by free energy minimization procedure of director field and finite-difference time-domain method, respectively. The domains, periodic structures and two-dimensional defects proper to the LC cells with tangential-conical anchoring have been studied depending on the layer thickness and cholesteric pitch.

Keywords: liquid crystal; cholesteric; nematic; conical boundary conditions; orientational structure; director configuration; topological defect

1. Introduction

Cholesteric liquid crystals (CLCs) are characterized by the helicoidal structure of director **n** (the unit vector indicating the preferred orientation of the long axes of liquid crystal (LC) molecules). These media have unique structural and optical properties [1]. CLCs can be used in various applications such as electro-optical devices with memory effect [2,3], quantum generation of light [4,5], switchable diffraction gratings [6–10], optical rotators [11], the formation of colloidal systems with a periodic distribution of particles [12], etc. The applications require various stable and metastable orientational structures depending on the boundary conditions, LC material parameters, ratio between LC layer thickness d and cholesteric pitch p, external factors [1,13–16]. For example, the homeotropic director orientation is formed in the cells at normal CLC anchoring with substrates and confinement ratio $d/p < K_{33}/2K_{22}$, where K_{22}, K_{33} are the twist and bend elastic constants, respectively [17–19]. The electric field applied to such cells causes the formation of bubble or elongated domains if the cholesteric pitch is slightly larger than the thickness of the LC layer [20,21]. Besides, the topological soliton-like structures can be obtained using laser radiation [22,23], the thermal quenching process [24] or the specific treatment of the orienting surface [25]. At tangential anchoring (non-degenerate planar anchoring) of CLC with the substrates, the Grandjean planar texture or domain structures are formed [3,13,26,27]. The structure with a periodic director modulation can be obtained in the LC cells with certain confinement ratios d/p by application of electric field [7,28]. The structure of modulated hybrid-aligned cholesteric is formed under hybrid surface anchoring (one substrate specifies the tangential boundary conditions and another substrate specifies the normal ones) in the cell with confinement ratio $d/p > 1$ and can be controlled by light [9], temperature [10] or electric field [29].

In the case of conical surface anchoring the director is tilted to the substrate and the azimuthal direction is degenerated (Figure 1). The structures of cholesteric with conical boundary conditions have not been well studied yet. Smooth transformations of cholesteric orientational structures induced by the modification of normal surface anchoring to tangential one through the formation of tilted or conical boundary conditions have been observed in [30,31]. A modulated structure of cholesteric under tangential surface anchoring at one of the substrates and weak conical boundary conditions formed at the interface of CLC and its isotropic phase at the opposite side have been investigated in [32]. It has been shown that the period of director modulation is equal to the cholesteric pitch and the orientation of periodic structure depends linearly on the confinement ratio d/p. At that, the critical d/p value for formation of the modulated structure depends on the ratio of elastic constants and it was less than in the case of normal or hybrid surface anchoring for some CLC materials.

In present work, the nematic and cholesteric structures formed in the LC cells with tangential-conical boundary conditions have been investigated.

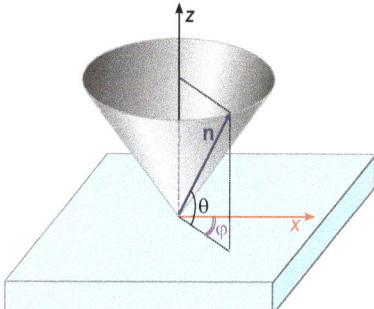

Figure 1. Scheme of conical surface anchoring of director **n** with the substrate. θ and φ are the tilt and azimuthal angles, respectively.

2. Materials and Methods

2.1. Experimental

The experiment was carried out with sandwich-like cells consisting of two glass substrates coated with polymer films and the LC layer between them. Bottom substrate was covered by the polyvinyl alcohol (PVA) (Sigma Aldrich, St. Louis, MO, USA) and the top one was covered by the poly(isobutyl methacrylate) (PiBMA) (Sigma Aldrich, St. Louis, MO, USA). The polymer films were deposited on the substrates by spin coating. The PVA film was unidirectionally rubbed while the PiBMA film was not treated after the deposition process. The LC layer thickness d assigned by the glass microspheres was measured by means of the interference method with spectrometer HR4000 (Ocean Optics, Largo, FL, USA) before the filling process. The nematic mixture LN-396 (Belarusian State Technological University, Minsk, Belarus) and LN-396 doped with the left-handed chiral additive cholesterylacetate (Sigma Aldrich, St. Louis, MO, USA) were used as a nematic and cholesteric, respectively. For the nematic mixture LN-396 the PiBMA film specifies the conical boundary conditions with the tilt angle 50° [33–35] and azimuthal degeneration. The polar anchoring strength of LN-396 on the PiBMA film is $W_0 \sim 10^{-6}$ J/m² [33]. The cells were filled by the LC in the mesophase at room temperature. After the filling process, the cells were kept for at least 24 h before measurements. The helical twisting power HTP = 6.9 µm^{-1} of cholesterylacetate in the LN-396 was determined using the Grandjean-Cano method. The cholesteric pitch p of used mixtures was calculated from $p = 1/(HTP \times c_w)$, where c_w is the weight concentration of the chiral additive. The LC cells with the confinement ratios d/p equal to 0.14, 0.28, 0.43, 0.60, 0.78 and 0.88 were investigated by means of the polarizing optical microscope (POM)

Axio Imager.A1m (Carl Zeiss, Oberkochen, Germany). The experiments were carried out using white light or quasi-monochromatic light with wavelength $\lambda = 602$ nm produced by the interference filter.

2.2. Computer Simulations

The nematic orientational structure within the layer was calculated by means of the free energy minimization [36]. The method is widely used for the LC systems without singularities of the director field. The Frank elastic energy density F_k was expressed as:

$$F_k = \frac{1}{2}k_{11}(\nabla \cdot \mathbf{n})^2 + \frac{1}{2}k_{22}(\mathbf{n} \cdot \nabla \times \mathbf{n} + 2\pi/p)^2 + \frac{1}{2}k_{33}(\mathbf{n} \times \nabla \times \mathbf{n})^2,$$

here $\mathbf{n}$ is the director, k_{11}, k_{22} and k_{33} are the splay, twist and bend elasticity coefficients, respectively. The main part of LN-396 is the alkyl-cyanobiphenyl and alkyloxy-cyanobiphenyls [33]. Therefore, it can be supposed that the elastic constants of LN-396 and E7 have close values. In the simulations we have taken $k_{11} = 11.1$ pN, $k_{22} = 7.6$ pN, $k_{33} = 17.1$ pN [37,38].

This function was converted into a discrete array of numbers, corresponding to thin, semihomogeneous sublayers. Each layer elastic energy was minimized in gradient direction. The five gradient components can be interpreted as elastic torques to rotate director and find θ, ϕ pair. In this study, we used 100 sublayers for a fairly accurate solution. Numerical convergence of this function was controlled by artificial viscosity. The domain wall orientational structure was analytically interpolated between the neighbouring domain structures. The asymptotic lateral boundary conditions were used. The values of $\theta(x)$ were reversed approaching the domain bulk value with multiplication by $\tan^{-1}(4x/L) \cdot 2/\pi$, where L is a characteristic wall thickness. The values of $\varphi(x)$ were adjusted by $\varphi_0 \cosh^{-1}(4x/L)$, where φ_0 were taken from experimental data for each domain wall.

The optical properties of LC structures were calculated by Finite-Difference Time-Domain (FDTD) method in commercial Lumerical package. LC structure was illuminated from below by the plane wave with normal incidence along the z-axis and polarization perpendicular to the rubbing direction. The plane wave source was located at the bottom boundary of the LC layer. Periodic boundary conditions were applied at the lateral boundaries of the simulation box (along the x and y axes), while the perfectly matched layers (PML) were used on the remaining top and bottom sides. The components of the electric field were calculated at the top boundary of the LC layer. POM images were obtained for the wavelength of incident light $\lambda = 602$ nm.

3. Results and Discussion

3.1. Nematic Layer

PVA film orients the nematic LN-396 tangentially while PiBMA film specifies the conical boundary conditions with the tilt angle of director $\theta_d = 50°$ [33–35] and azimuthal degeneration. The azimuthal degeneration is eliminated by presence of rubbed PVA film on the bottom substrate specifying the director orientation with azimuthal angle $\varphi_0 = 0°$. As a result, the orientational structure with zero azimuthal angle of the director is formed in the cells including the LC-PiBMA interface (Figure 2). At that, two types of domains differing by the tilt angles $+\theta_d$ and $-\theta_d$ on the PiBMA film are observed. The similar situation was observed in the LC cell with tangential-conical boundary conditions in the case of conical surface anchoring at the interface of LC-isotropic phase [39]. Domains are separated from each other by defect walls which are clearly observed on the optical textures using polarized light (without the crossed analyzer) when the polarization of incident light coincides with rubbing direction of PVA film. In the case of orthogonal orientation of polarizer and rubbing direction of PVA film these defect walls are almost invisible (see Supplementary Materials Figure S1). When the LC cell is placed between crossed polarizers the segments of wall with orientation differing from rubbing direction are clearly observed (Figure 2a) indicating the presence of twist deformation of the director. The similar situation was shown in the domain walls formed by the application of the magnetic field [40]. The director possesses the homogeneous azimuthal orientation along the rubbing direction far from the domain walls (Figure 2a) in the LC cells filled by the nematic in the mesophase and it can be considered as the evidence of the strong anchoring of LN-396 at the PVA film. The similar optical textures were observed in various regions of the LC cell showing the uniformity of boundary conditions across the surfaces of substrates.

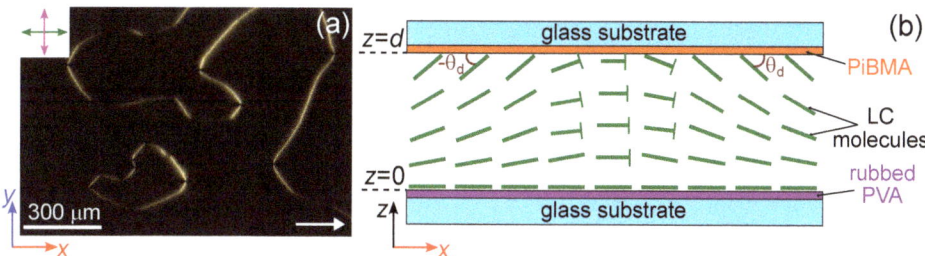

Figure 2. POM photo of LN-396 nematic layer with PiBMA film on the top substrate ($z = d$) and rubbed PVA film on the bottom one ($z = 0$) (**a**); The scheme of LC orientation in domains with tilt angle of director $+\theta_d$ and $-\theta_d$ on the PiBMA film (**b**). The LC layer thickness is $d = 21$ μm. Hereinafter, the polarizer and analyzer directions are indicated by the magenta and green double arrows, respectively. The single arrow is the rubbing direction of PVA film.

The azimuthal angle of director rotation φ_d on the top substrate covered with PiBMA film can be measured by means of the analyzer rotation (Figure 3). If the polarization of incident light is parallel or orthogonal to the director on the entrance substrate covered with PVA and Mauguin condition [41] is valid then the polarization of light passed through the LC cell remains linear but rotated relative to the incident light polarization. The angle of polarization rotation is equal to the azimuthal angle of director rotation. In the case of the orthogonal orientation of the light polarization and the director on the entrance substrate the light passed through the LC cell is polarized perpendicularly to the director projection on the plane of the output substrate. Consequently, the darkest areas of optical texture correspond to the parallel orientation of analyzer and director projection on the plane of output substrate covered with PiBMA film. Thus, the topology of the director projection over the area of the substrate covered with PiBMA can be determined.

The optical textures of nematic layer for different β (angle between analyzer and rubbing direction of PVA film) and the director orientation on the PiBMA film near the domain wall are presented in Figure 3. The analysis by the above-mentioned technique revealed that there is the smooth azimuthal rotation of director from $\varphi_d = 0$ (far away from the wall) to the value φ_d in the center of the wall where the director is parallel to its plane. The rotation occurs over a distance of approximately the LC layer thickness. The domain wall contains the reversing points [42], dividing the segments of the wall with different director orientation relative to rubbing direction. The direction of azimuthal rotation (the φ sign) is defined by the condition that the absolute value of azimuthal angle of rotation on the substrate covered with PiBMA does not exceed $\pi/2$ value.

The cross-section of the domain wall is shown in Figure 4a. The angle between the wall plane and rubbing direction is 60°. The orientational structure was calculated by the minimization of elastic energy and analytical interpolation of director between two neighboring domains. The POM photos of domain wall segment for the different β angles have been obtained by the FDTD method (Figure 4b) based on the calculated orientational structure. When the analyzer is parallel to the rubbing direction ($\beta = 0°$) the darkest area is observed far away from the wall. At the same time, the center of domain wall is bright. The variation of β angle leads to the appearance of a couple of dark areas (extinction bands) merging at the center line of the wall at $\beta = -60°$ corresponded to the parallel orientation of the analyzer and the wall. The director orientation near the domain wall on the substrate with conical boundary conditions ($z = d$) is shown in Figure 4c. The dependence of the azimuthal angle of director rotation φ_d on the distance h from the center of the domain wall has been obtained (Figure 4d, solid line). Besides, the experimentally measured dependence $\varphi_d(h)$ using the analyzer rotation method are also shown in Figure 4d. The calculated function $\varphi_d(h)$ is in a good agreement with the experimental data.

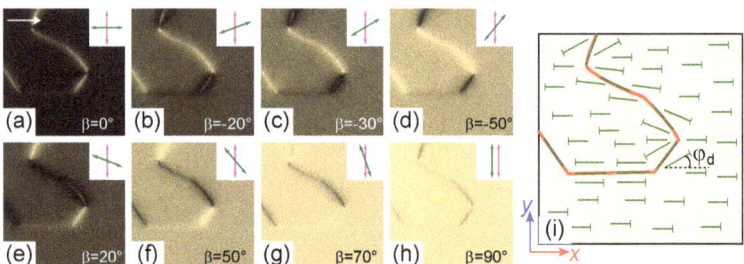

Figure 3. POM photos of sample area presented in Figure 2. The polarizer is perpendicular to the rubbing direction of PVA film. The β angle between the analyzer and rubbing direction is 0° (**a**), −20° (**b**), −30° (**c**), −50° (**d**), 20° (**e**), 50° (**f**), 70° (**g**), 90° (**h**). Corresponding scheme of director orientation at the top substrate covered with PiBMA (**i**). The POM photos show 380 × 380 µm area. The domain wall is indicated by the red line. The angle φ_d is an azimuthal angle of director at the top substrate.

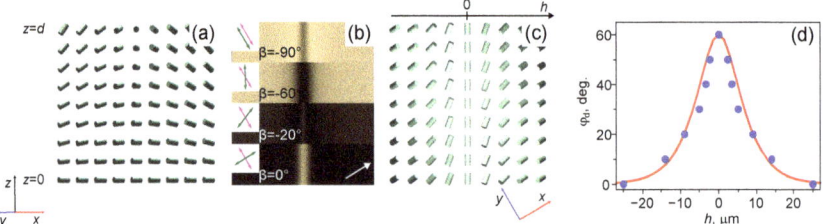

Figure 4. The calculated director configuration of a nematic in the cross-section of the layer near the domain wall (**a**); The image of segment of domain wall for different β angles between analyzer and rubbing direction obtained by the FDTD method (**b**); The director orientation near the domain wall on the substrate covered with PiBMA film (**c**); The theoretically calculated (solid line) and experimentally measured (points) dependences of the azimuthal angle of director rotation φ_d on the distance h from the center of the domain wall (**d**). The LC layer thickness is $d = 21$ µm.

3.2. Twisted Cholesteric Structure

In the nematic LC cells, the defect walls separating the areas with the opposite values of the tilt angle θ_d on the PiBMA film were observed (Figures 2a and 3). At that, the different segments of the wall had the opposite sign of the azimuthal director rotation and these segments were separated from each other by the reversing points. In the cholesteric LC cells, the right-handed segments of the walls are less stable that increases the monodomain sizes. POM photos of samples with the LC layer thickness $d = 6.5$ µm and confinement ratios $d/p = 0.14$, 0.28 are shown in Figure 5a,b. One can see that the monodomains corresponding to the opposite values of tilt angle θ_d in the cell with confinement ratio $d/p = 0.14$ have a larger size (Figure 5a) than in the nematic LC cells (Figure 2a). The monodomain structure is formed in the cell with confinement ratio $d/p = 0.28$ (Figure 5b).

In the general case, the Mauguin condition is not valid for the homogeneously twisted structure of LC. However, the layer thickness of LC and wavelength of incident light can be chosen so that the Gooch-Terry minimum condition is valid [43]. In this case, the polarization of light passed through the LC cell remains linear but rotated relative to the initial state if the linear polarization of incident light is parallel or orthogonal to the director on the entrance substrate. The angle of polarization rotation is equal to the azimuthal angle of director rotation. The Gooch-Terry minimum condition is independent of the rotation angle of director [41]. The twisted structure with azimuthal angle of director rotation depending on the confinement ratio d/p is formed in the cholesteric cell with tangential-conical boundary conditions (Figure 5e). The tilt and azimuthal angles of the director in the cholesteric bulk were calculated by the minimization of elastic energy for three d/p ($d = 6.5\,\mu m$) values (Figure 5e). The ellipticity angle ε for polarized light with wavelength $\lambda = 602\,nm$ was determined based on these data (Figure 5f). One can see that Gooch-Terry minimum condition is nearly independent of the confinement ratio d/p as in the case of the homogeneously twisted LC structure. For the LC layer thickness $d = 6.5\,\mu m$, the wavelength of incident light $\lambda = 602\,nm$ nearly corresponds to the Gooch-Terry minimum condition and the ε value do not exceed 5° after passing the linearly polarized light through the cells with $d/p = 0.14$, 0.28 and 0.44. Thus, the method of analyzer rotation can be used for experimental measuring the azimuthal angle φ_d of the director on the top substrate covered with PiBMA.

The darkest state of domains for the LC cell with $d/p = 0.14$ is observed at $\beta = 37°$ when the interference filter ($\lambda = 602\,nm$) is used (Figure 5c). Thus, the azimuthal angle of director rotation at the substrate covered with PiBMA is $\varphi_d \cong 37°$. For the LC cell with $d/p = 0.28$ the darkest state of domain (for incident light with wavelength $\lambda = 602\,nm$) is observed at $\beta = 75°$ (Figure 5d) and, consequently, $\varphi_d \cong 75°$.

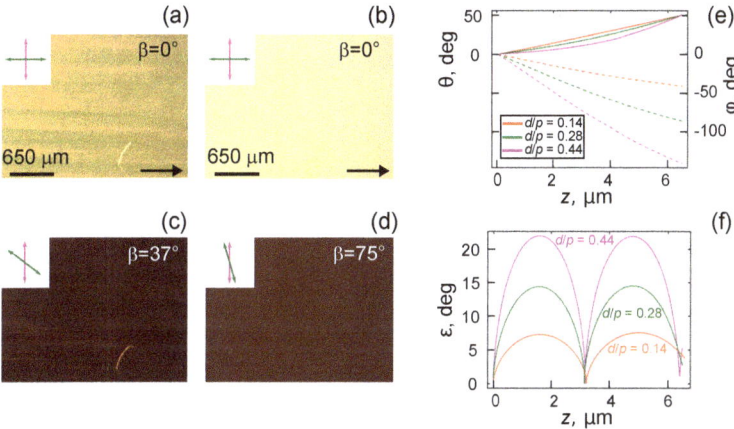

Figure 5. POM photos of cholesteric layers with confinement ratio $d/p = 0.14$ (**a**,**c**) and $d/p = 0.28$ (**b**,**d**). Calculated dependences of tilt angle $\theta(z)$ (solid lines), azimuthal angle $\varphi(z)$ (dashed lines) (**e**) and corresponding dependences of ellipticity angle $\varepsilon(z)$ (**f**) for the samples with $d/p = 0.14$ (orange line), $d/p = 0.28$ (green line) and $d/p = 0.44$ (magenta line). POM photos are taken using white light when β angle between analyzer and rubbing direction of PVA film is 0° (**a**,**b**). POM photos are taken using the interference filter ($\lambda = 602\,nm$) when β angle is 37° (**c**) and 75° (**d**). Polarizer is perpendicular to the rubbing direction. LC layer thicknesses are 6.5 μm.

Structure deformations are observed in the LC cell with $d = 6.5\,\mu m$ and confinement ratio $d/p = 0.44$. In this case, the defects are observed in the form of elongated loops (Figure 6a) or lines originating and ending at the cell edges (interfaces LC-air and LC-glue) or the surface defects. Therefore, the defects number near the edges of LC cell is more than in the central area. They are well observed independently of the polarization of incident light when the LC cell is placed between crossed polarizers. Under analyzer switched-off the lines are clearly visible when the polarizer is parallel to the rubbing direction and the lines are almost invisible when the polarization of incident light is perpendicular to the rubbing direction. The orientation of loops and their relative position can be different. At the same time, there is the twisted director structure between loops and the optical texture has the darkest state (for incident light with wavelength $\lambda = 602$ nm) at $\beta \cong -40°$ (Figure 6b), corresponding to the azimuthal angle of director rotation $\varphi_d \cong 140°$. The theoretically calculated φ_d values (Figure 5e) are in a good agreement with the experimentally measured φ_d values (Table 1).

Table 1. The theoretically calculated (φ_d^{calc}) and experimentally measured (φ_d^{exp}) values of the azimuthal angle of director rotation at the substrate covered with PiBMA in the LC cells with $d/p = 0.14, 0.28, 0.44$.

d/p	φ_d^{calc}	φ_d^{exp}
0.14	41.3°	37° ± 3°
0.28	80.3°	75° ± 3°
0.44	140.0°	140° ± 5°

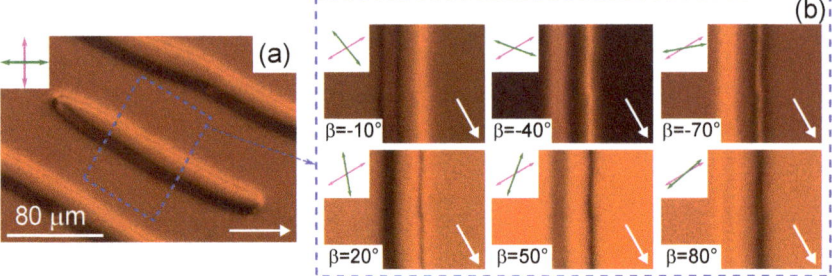

Figure 6. POM photos of cholesteric layer with confinement ratio $d/p = 0.44$ are taken using the interference filter ($\lambda = 602$ nm). The angle β between analyzer and rubbing direction of PVA film is 0° (**a**); The magnified area of 90 × 65 µm of loop at different β angles (**b**). Polarizer is perpendicular to the rubbing direction. LC layer thickness is 6.5 µm.

A couple of extinction bands near one of the loop lines can be observed (Figure 6b) by the analyzer rotation. As in the case of nematic described above the couple of extinction bands are gathered to the defect under the analyzer rotation. These lines are merged when the analyzer is almost parallel to the considered segment of the loop. Under further analyzer rotation the similar couple of extinction lines appears near the opposite segment of the loop and the distance between these lines depends on the β angle (Figure 6b). Such orientation of extinction lines indicates that the director is oriented parallel to the line of loop defect on the top substrate and π azimuthal rotation angle of director occurs between two opposite segments of the loop. Thus, the values of azimuthal angles φ_d at the opposite segments of the loop differ by π and the tilt angle θ_d is zero at the defect line. For example, the angle between the loop line and rubbing direction is approximately 30° (Figure 6a). Consequently, the rotation angle φ_d is about 210° at the first loop segment and 30° at the opposite one.

The orientation with different angle of director rotation was simulated by the FDTD method (Figure 7). The calculation was performed based on the spatial distribution of director in the cholesteric layer determined by the elastic energy minimization as in case of nematic. The section of defect loop by the plane perpendicular to the defect lines and oriented at 30° to the rubbing direction is shown in Figure 7a. POM images of cholesteric layer with $d/p = 0.44$ for different β angles between analyzer and rubbing direction were calculated (Figure 7b). It was revealed that the darkest state of optical texture corresponds to the $\beta = -40°$ far away from the defect. At the same time, the brightest optical texture is observed at $\beta = 50°$. The defect line corresponding to the smaller twist angle becomes dark when the analyzer is parallel to this line, i.e., it is parallel to the director on the top substrate. At that, the second defect line becomes dark when the angle between analyzer and wall is approximately 10°. It is explained by the fact that the effective refractive index near the walls has the larger value than that one far away from them. In this case the wavelength of incident light $\lambda = 602$ nm does not satisfy the Gooch-Terry minimum condition at large azimuthal twist of the director. As a consequence, the defect line with a large twist angle has the darkest state when the analyzer is not parallel to the director on the top substrate but this line is appreciably brighter than other sample areas at the minimum intensity condition of transmitted light. Spatial distribution of director on the substrate with conical boundary conditions is shown in Figure 7c. The dependence of the azimuthal angle of director rotation φ_d on the distance h from the center of the left domain line in Figure 7c has been obtained (Figure 7d, solid line). Besides, the experimentally measured dependence $\varphi_d(h)$ using the analyzer rotation method are also shown in Figure 7d. The calculated function $\varphi_d(h)$ is in a good agreement with the experimental data.

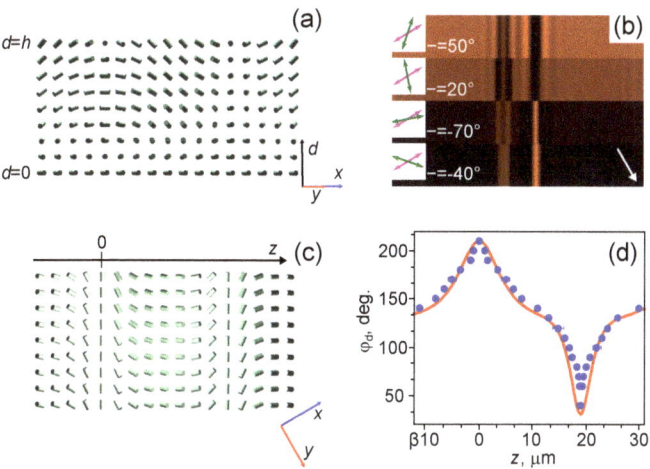

Figure 7. Calculated director configuration in the section perpendicular to the cell substrates and lines of defect (**a**); Simulated POM image of the area with couple of defects at different values of β angle between the analyzer and rubbing direction (**b**); The director orientation near the defect on the substrate covered with PiBMA (**c**); The theoretically calculated (solid line) and experimentally measured (points) dependences of the azimuthal angle of director rotation φ_d on the distance h from the center of the left domain line (**d**). The polarizer is perpendicular to the rubbing direction on the simulated POM image. LC layer thickness is 6.5 µm and confinement ratio is $d/p = 0.44$.

3.3. Periodic Structure

Quasi-periodic structure of defects is formed in the LC cell with $d = 6.5$ µm and confinement ratio $d/p = 0.44$ at the area close to the cell edge (Figure 8a). The azimuthal angle of director rotation φ_d on the defect lines differs by π ($-\pi$) and between them φ_d has intermediate values (see Supplementary Figure S2). Similar periodic structure is observed through the whole cell area with $d/p = 0.60$ and in the samples with larger values of confinement ratio (Figure 8b–d). The period of observed structure is about $2p$. The periodic structure was formed at similar values of confinement ratios d/p in the LC cells with tangential-weak conical boundary conditions [32], but in that work the period was close to the cholesteric pitch. The disturbance of periodicity of two types can be observed in the samples (Figure 8e). In the first case, the additional couple of defect lines appears leading to the local increasing of period near the region of line bending. In this area one defect line with larger azimuthal twist angle is smoothly transformed into the defect line with a smaller φ_d angle. In the second case, some defect lines are formed and their orientation differs from the majority of defect lines. For example, some defect lines are oriented vertically in Figure 8c. The periodic defect lines have 180° turn near these lines and do not intersect them. This way the defect line with larger azimuthal angle of director rotation is modified into line with smaller φ_d angle and vice versa.

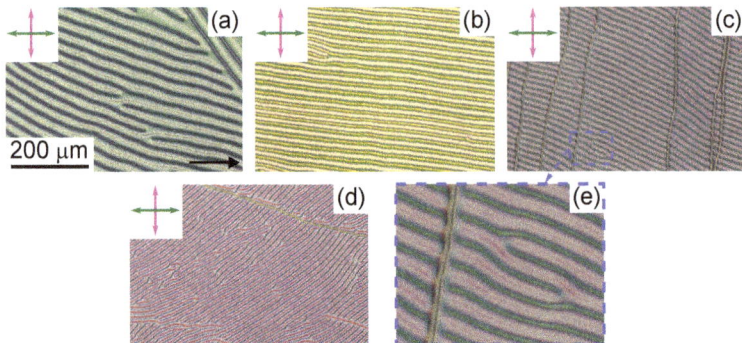

Figure 8. POM photos of cholesteric layers with confinement ratios d/p: 0.44 (**a**), 0.60 (**b**), 0.78 (**c**) and 0.88 (**d**). The magnified area of 100×70 µm with two types of defects of periodic structure at $d/p = 0.78$ (**e**). POM photos (**a**–**d**) have the same scale. LC layer thicknesses are 6.5 µm.

It should be noted that the formation of defects and their periodic structure depends not only on the confinement ratio d/p but also on the value of cholesteric pitch (the thickness of LC layer). For example, the periodic structure is formed in the LC cell with $d/p = 0.60$ at $d = 6.5$ µm ($p = 10.5$ µm). But the twisted structure similar to the structure observed in LC cell with $d = 6.5$ µm and $d/p = 0.44$ (Figures 6 and 8a) is formed in the sample with LC layer thickness $d = 13$ µm ($p = 21$ µm). The twisted structure with 185° azimuthal angle of director rotation containing a small number of defects is formed in the LC cell at $d = 21$ µm ($p = 37$ µm). The observed dependence of director configuration on the value of cholesteric pitch (the thickness of LC layer) in LC cells with the same ratio d/p is probably caused by the conical boundary condition on the top substrate. In the system under study the polar anchoring strength of LN-396 on PiBMA is not strong and the tilt angle on this substrate is changed near the defect loop.

4. Conclusions

The orientational structures of nematic and cholesteric LC with the tangential-conical boundary conditions were investigated. The cells with various LC layer thickness d and confinement ratio d/p have been considered. The domain structure was formed in the nematic LC cells containing the areas of positive and negative tilt angle of director at the substrate with conical boundary conditions. The domains were divided by the walls where the tilt angle of director was 0° at both substrates. The monodomain twisted structure without defects was formed in the cholesteric LC cell with $d = 6.5$ µm and $d/p = 0.28$. At that, the azimuthal angle of director rotation was $\varphi_d = 75°$ at the substrate with conical boundary conditions. Increase of the confinement ratio d/p up to 0.44 led to the appearance of elongated loop defects. These topological defects were characterized by the difference of azimuthal angle of director rotation by $\pm \pi$ at the opposite segments of loop and 0° tilt angle of director at the defect. The periodic structure with a period close to the double cholesteric pitch was formed at $d/p \geq 0.60$. The threshold value of d/p to form the periodic structure was relatively small. At the same time, this confinement ratio depended on the LC layer thickness (cholesteric pitch) in contrast to normal boundary conditions [17–19]. It was probably caused by the features of asymmetric boundary conditions in the cell.

It was demonstrated that the azimuthal angle φ_d of director at the substrate covered with PiBMA depends on the cholesteric pitch p. Consequently, the φ_d can be tuned using the different external factors (electric field, temperature, light radiation, etc.) changing p and it is interesting to future research. Besides, the described periodic structures formed in the cholesteric LC cells (Figure 8) have satisfactory quality and can be proposed for the various applications as diffraction gratings. One of the advantages of the CLC diffraction gratings is their tunability. Thus, it is important to investigate the diffraction patterns and their transformations upon application of the external factors to the LC cell.

Supplementary Materials: The following are available at http://www.mdpi.com/2073-4352/9/5/249/s1, Figure S1. POM photos of nematic layer taken using polarized light. The angle between rubbing direction of PVA film and polarizer is 0° (**a**), 45° (**b**) and 90° (**c**). The polarizer direction is indicated by the double arrow and the single arrow is the rubbing direction of PVA film, Figure S2. POM photos of cholesteric layer with confinement ratio d/p = 0.44 are taken using the interference filter (l = 602 nm). The angle between analyzer and rubbing direction of PVA film is −10° (**a**), −40° (**b**), −70° (**c**), 20° (**d**), 50° (**e**), 70° (**f**). Polarizer is perpendicular to the rubbing direction. LC layer thickness is 6.5 mm.

Author Contributions: M.N.K. initiated this study; V.S.S. and M.N.K. performed the experiments and analysed the optical patterns, R.G.B. and I.V.T. performed a simulation of the orientation structures and optical texture, M.N.K. and V.Y.Z. supervised the study. All authors wrote and reviewed the manuscript.

Funding: This work was supported by the Russian Science Foundation (No. 18-72-10036).

Conflicts of Interest: The authors declare no conflict of interest.

References

1. Oswald, P.; Pieranski, P. *Nematic and Cholesteric Liquid Crystals: Concepts and Physical Properties Illustrated by Experiments*; The Liquid Crystals Book Series; Taylor & Francis: Boca Raton, FL, USA, 2005.
2. Kim, J.H.; Huh, J.W.; Oh, S.W.; Ji, S.M.; Jo, Y.S.; Yu, B.H.; Yoon, T.H. Bistable switching between homeotropic and focal-conic states in an ion-doped chiral nematic liquid crystal cell. *Opt. Express* **2017**, *25*, 29180–29188, doi:10.1364/OE.25.029180. [CrossRef]
3. Hsiao, Y.C.; Tang, C.Y.; Lee, W. Fast-switching bistable cholesteric intensity modulator. *Opt. Express* **2011**, *19*, 9744–9749, doi:10.1364/OE.19.009744. [CrossRef] [PubMed]
4. Il'chishin, I.P.; Tikhonov, E.A.; Tishchenko, V.G.; Shpak, M.T. Generation of tunable radiation by impurity cholesteric liquid crystals. *JETP Lett.* **1981**, *32*, 24–27.
5. Kopp, V.I.; Fan, B.; Vithana, H.K.M.; Genack, A.Z. Low-threshold lasing at the edge of a photonic stop band in cholesteric liquid crystals. *Opt. Lett.* **1998**, *23*, 1707–1709, doi:10.1364/OL.23.001707. [CrossRef] [PubMed]
6. Subacius, D.; Bos, P.J.; Lavrentovich, O.D. Switchable diffractive cholesteric gratings. *Appl. Phys. Lett.* **1997**, *71*, 1350–1352, doi:10.1063/1.119890. [CrossRef]
7. Subacius, D.; Shiyanovskii, S.V.; Bos, P.; Lavrentovich, O.D. Cholesteric gratings with field-controlled period. *Appl. Phys. Lett.* **1997**, *71*, 3323–3325, doi:10.1063/1.120325. [CrossRef]
8. Senyuk, B.I.; Smalyukh, I.I.; Lavrentovich, O.D. Switchable two-dimensional gratings based on field-induced layer undulations in cholesteric liquid crystals. *Opt. Lett.* **2005**, *30*, 349, doi:10.1364/OL.30.000349. [CrossRef]
9. Ryabchun, A.; Bobrovsky, A.; Stumpe, J.; Shibaev, V. Rotatable Diffraction Gratings Based on Cholesteric Liquid Crystals with Phototunable Helix Pitch. *Adv. Opt. Mater.* **2015**, *3*, 1273–1279, doi:10.1002/adom.201500159. [CrossRef]
10. Lin, C.H.; Chiang, R.H.; Liu, S.H.; Kuo, C.T.; Huang, C.Y. Rotatable diffractive gratings based on hybrid-aligned cholesteric liquid crystals. *Opt. Express* **2012**, *20*, 26837, doi:10.1364/OE.20.026837. [CrossRef] [PubMed]
11. Liu, C.K.; Chiu, C.Y.; Morris, S.M.; Tsai, M.C.; Chen, C.C.; Cheng, K.T. Optically Controllable Linear-Polarization Rotator Using Chiral-Azobenzene-Doped Liquid Crystals. *Materials* **2017**, *10*, 1299, doi:10.3390/ma10111299. [CrossRef]
12. Varney, M.C.M.; Zhang, Q.; Senyuk, B.; Smalyukh, I.I. Self-assembly of colloidal particles in deformation landscapes of electrically driven layer undulations in cholesteric liquid crystals. *Phys. Rev. E* **2016**, *94*, doi:10.1103/PhysRevE.94.042709. [CrossRef]
13. Dierking, I. *Textures of Liquid Crystals*; Wiley-VCH: Weinheim, Germany, 2003.
14. Ma, L.L.; Li, S.S.; Li, W.S.; Ji, W.; Luo, B.; Zheng, Z.G.; Cai, Z.P.; Chigrinov, V.; Lu, Y.Q.; Hu, W.; et al. Rationally Designed Dynamic Superstructures Enabled by Photoaligning Cholesteric Liquid Crystals. *Adv. Opt. Mater.* **2015**, *3*, 1691–1696, doi:10.1002/adom.201500403. [CrossRef]
15. Zheng, Z.G.; Li, Y.; Bisoyi, H.K.; Wang, L.; Bunning, T.J.; Li, Q. Three-dimensional control of the helical axis of a chiral nematic liquid crystal by light. *Nature* **2016**, *531*, 352–356, doi:10.1038/nature17141. [CrossRef] [PubMed]
16. Nys, I.; Chen, K.; Beeckman, J.; Neyts, K. Periodic Planar-Homeotropic Anchoring Realized by Photoalignment for Stabilization of Chiral Superstructures. *Adv. Opt. Mater.* **2018**, *6*, 1701163, doi:10.1002/adom.201701163. [CrossRef]

17. Zel'dovich, B.Y.; Tabiryan, N.V. Freedericksz transition in cholesteric liquid crystals without external fields. *JETP Lett.* **1981**, *34*, 406–408.
18. Cladis, P.E.; Kléman, M. The Cholesteric Domain Texture. *Mol. Cryst. Liq. Cryst.* **1972**, *16*, 1–20, doi:10.1080/15421407208083575. [CrossRef]
19. Goossens, W.J.A. The influence of homeotropic and planar boundary conditions on the field induced cholesteric-nematic transition. *J. Phys.* **1982**, *43*, 1469–1474, doi:10.1051/jphys:0198200430100146900. [CrossRef]
20. Oswald, P.; Baudry, J.; Pirkl, S. Static and dynamic properties of cholesteric fingers in electric field. *Phys. Rep.* **2000**, *337*, 67–96, doi:10.1016/S0370-1573(00)00056-9. [CrossRef]
21. Varanytsia, A.; Posnjak, G.; Mur, U.; Joshi, V.; Darrah, K.; Muševič, I.; Čopar, S.; Chien, L.C. Topology-commanded optical properties of bistable electric-field-induced torons in cholesteric bubble domains. *Sci. Rep.* **2017**, *7*, doi:10.1038/s41598-017-16241-4. [CrossRef]
22. Ackerman, P.J.; Qi, Z.; Smalyukh, I.I. Optical generation of crystalline, quasicrystalline, and arbitrary arrays of torons in confined cholesteric liquid crystals for patterning of optical vortices in laser beams. *Phys. Rev. E* **2012**, *86*, 021703, doi:10.1103/PhysRevE.86.021703. [CrossRef]
23. Ackerman, P.J.; Trivedi, R.P.; Senyuk, B.; van de Lagemaat, J.; Smalyukh, I.I. Two-dimensional skyrmions and other solitonic structures in confinement-frustrated chiral nematics. *Phys. Rev. E* **2014**, *90*, 012505, doi:10.1103/PhysRevE.90.012505. [CrossRef] [PubMed]
24. Kim, Y.H.; Gim, M.J.; Jung, H.T.; Yoon, D.K. Periodic arrays of liquid crystalline torons in microchannels. *RSC Adv.* **2015**, *5*, 19279–19283, doi:10.1039/C4RA16883F. [CrossRef]
25. Nys, I.; Beeckman, J.; Neyts, K. Surface-Mediated Alignment of Long Pitch Chiral Nematic Liquid Crystal Structures. *Adv. Opt. Mater.* **2018**, *6*, 1800070, doi:10.1002/adom.201800070. [CrossRef]
26. Andrienko, D. Introduction to liquid crystals. *J. Mol. Liq.* **2018**, *267*, 520–541, doi:10.1016/j.molliq.2018.01.175. [CrossRef]
27. Belyaev, S.V.; Rumyantsev, V.G.; Belyaev, V.V. Optical and electro-optical properties of confocal cholesteric textures. *JETP* **1977**, *46*, 337–340.
28. Belyaev, S.V.; Blinov, L.M. Instability of planar texture of a cholesteric liquid crystal in an electric field. *JETP* **1976**, *43*, 96–99.
29. Nose, T.; Miyanishi, T.; Aizawa, Y.; Ito, R.; Honma, M. Rotational Behavior of Stripe Domains Appearing in Hybrid Aligned Chiral Nematic Liquid Crystal Cells. *Jpn. J. Appl. Phys.* **2010**, *49*, 051701, doi:10.1143/JJAP.49.051701. [CrossRef]
30. Kumar, T.A.; Le, K.V.; Aya, S.; Kang, S.; Araoka, F.; Ishikawa, K.; Dhara, S.; Takezoe, H. Anchoring transition in a nematic liquid crystal doped with chiral agents. *Phase Transit.* **2012**, *85*, 888–899, doi:10.1080/01411594.2012.692092. [CrossRef]
31. Tran, L.; Lavrentovich, M.O.; Durey, G.; Darmon, A.; Haase, M.F.; Li, N.; Lee, D.; Stebe, K.J.; Kamien, R.D.; Lopez-Leon, T. Change in Stripes for Cholesteric Shells via Anchoring in Moderation. *Phys. Rev. X* **2017**, *7*, 041029, doi:10.1103/PhysRevX.7.041029. [CrossRef]
32. Zola, R.S.; Evangelista, L.R.; Yang, Y.C.; Yang, D.K. Surface Induced Phase Separation and Pattern Formation at the Isotropic Interface in Chiral Nematic Liquid Crystals. *Phys. Rev. Lett.* **2013**, *110*, 057801, doi:10.1103/PhysRevLett.110.057801. [CrossRef] [PubMed]
33. Krakhalev, M.N.; Prishchepa, O.O.; Sutormin, V.S.; Zyryanov, V.Y. Director configurations in nematic droplets with tilted surface anchoring. *Liq. Cryst.* **2017**, *44*, 355–363, doi:10.1080/02678292.2016.1205225. [CrossRef]
34. Rudyak, V.Y.; Krakhalev, M.N.; Prishchepa, O.O.; Sutormin, V.S.; Emelyanenko, A.V.; Zyryanov, V.Y. Orientational structures in nematic droplets with conical boundary conditions. *JETP Lett.* **2017**, *106*, 384–389, doi:10.1134/S0021364017180102. [CrossRef]
35. Rudyak, V.Y.; Krakhalev, M.N.; Sutormin, V.S.; Prishchepa, O.O.; Zyryanov, V.Y.; Liu, J.H.; Emelyanenko, A.V.; Khokhlov, A.R. Electrically induced structure transition in nematic liquid crystal droplets with conical boundary conditions. *Phys. Rev. E* **2017**, *96*, 052701, doi:10.1103/PhysRevE.96.052701. [CrossRef]
36. Timofeev, I.V.; Lin, Y.T.; Gunyakov, V.A.; Myslivets, S.A.; Arkhipkin, V.G.; Vetrov, S.Y.; Lee, W.; Zyryanov, V.Y. Voltage-induced defect mode coupling in a one-dimensional photonic crystal with a twisted-nematic defect layer. *Phys. Rev. E* **2012**, *85*, 011705, doi:10.1103/PhysRevE.85.011705. [CrossRef] [PubMed]
37. *Merck E7 Liquid Crystal Datasheet*; Merck KGaA: Darmstadt, Germany.

38. Raynes, E.P.; Brown, C.V.; Strömer, J.F. Method for the measurement of the K22 nematic elastic constant. *Appl. Phys. Lett.* **2003**, *82*, 13–15, doi:10.1063/1.1534942. [CrossRef]
39. Vilanove, R.; Guyon, E.; Mitescu, C.; Pieranski, P. Mesure de la conductivité thermique et détermination de l'orientation des molécules a l'interface nématique isotrope de MBBA. *J. Phys.* **1974**, *35*, 153–162, doi:10.1051/jphys:01974003502015300. [CrossRef]
40. Gilli, J.; Morabito, M.; Frisch, T. Ising-Bloch transition in a nematic liquid crystal. *J. Phys. II* **1994**, *4*, 319–331, doi:10.1051/jp2:1994131. [CrossRef]
41. Yeh, P.; Gu, C. *Optics of Liquid Crystal Displays*; Wiley Series in Pure and Applied Optics; Wiley: New York, NY, USA, 1999.
42. Ryschenkow, G.; Kleman, M. Surface defects and structural transitions in very low anchoring energy nematic thin films. *J. Chem. Phys.* **1976**, *64*, 404–412, doi:10.1063/1.431934. [CrossRef]
43. Gooch, C.H.; Tarry, H.A. The optical properties of twisted nematic liquid crystal structures with twist angles ≤ 90 degrees. *J. Phys. D Appl. Phys.* **1975**, *8*, 1575–1584, doi:10.1088/0022-3727/8/13/020. [CrossRef]

© 2019 by the authors. Licensee MDPI, Basel, Switzerland. This article is an open access article distributed under the terms and conditions of the Creative Commons Attribution (CC BY) license (http://creativecommons.org/licenses/by/4.0/).

Article

Surface Anchoring Effects on the Formation of Two-Wavelength Surface Patterns in Chiral Liquid Crystals

Ziheng Wang, Pardis Rofouie and Alejandro D. Rey *

Department of Chemical Engineering, McGill University, 3610 University Street,
Montreal, QC H3A 2B2, Canada; ziheng.wang@mail.mcgill.ca (Z.W.); pardis.rofouieeraghi@mail.mcgill.ca (P.R.)
* Correspondence: alejadro.rey@mcgill.ca; Tel.: +1-514-398-4196

Received: 26 February 2019; Accepted: 26 March 2019; Published: 2 April 2019

Abstract: We present a theoretical analysis and linear scaling of two-wavelength surface nanostructures formed at the free surface of cholesteric liquid crystals (CLC). An anchoring model based on the capillary shape equation with the high order interaction of anisotropic interfacial tension is derived to elucidate the formation of the surface wrinkling. We showed that the main pattern-formation mechanism is originated due to the interaction between lower and higher order anchoring modes. A general phase diagram of the surface morphologies is presented in a parametric space of anchoring coefficients, and a set of anchoring modes and critical lines are defined to categorize the different types of surface patterns. To analyze the origin of surface reliefs, the correlation between surface energy and surface nano-wrinkles is investigated, and the symmetry and similarity between the energy and surface profile are identified. It is found that the surface wrinkling is driven by the director pressure and is annihilated by two induced capillary pressures. Linear approximation for the cases with sufficient small values of anchoring coefficients is used to realize the intrinsic properties and relations between the surface curvature and the capillary pressures. The contributions of capillary pressures on surface nano-wrinkling and the relations between the capillary vectors are also systematically investigated. These new findings establish a new approach for characterizing two-length scale surface wrinkling in CLCs, and can inspire the design of novel functional surface structures with the potential optical, friction, and thermal applications.

Keywords: cholesteric liquid crystal; two-length scale surface wrinkling; capillary shape equation; anisotropic surface energy

1. Introduction

A variety of periodic surface structures and wrinkled textures are widely found in the plant and animal kingdoms [1–6]. Since these surface ultrastructures with micro/nano scale features provide unique optical responses and iridescent colors [7–11], understanding their formation mechanism is crucial in realizing structural color in nature and in biomimetic design of novel photonic systems. As similar nano/micro scale periodic wrinkles are formed at the free surface of both synthetic and biological cholesteric liquid crystals (CLCs) [12,13], and CLC phases are widely found in Nature and living soft materials both in vivo and vitro [13,14], nematic liquid crystal self-assembly has been proposed as the formation mechanism of helicoidal plywoods and the surface ultrastructures in many fibrous composites ranging from plant cell walls to arthropod cuticles [15–19]. Moreover, it has been shown that the characteristics of chiral phases control the unique colors and optical properties exhibited in the films and fibers made by cellulose-based CLCs [20,21].

Inspired by surface ultrastructure in Nature, engineered surface structures incorporating chiral nematic structures can be fabricated to mimic the unique optical properties. If the formation of the surface patterns can be efficiently captured by a rigorous model based on a CLC mesophase, we can elucidate the pattern formation mechanisms for the construction of biomimetic proof-of-concept prototypes. In our previous works [22–24], significant efforts have been made in formulating and validating theoretical models to explain the formation of surface wrinkles in a plant-based CLC as a model material system. We identified the chiral capillary pressure, known as director pressure, that reflects the anisotropic nature of CLC through the orientation contribution to the surface energy as the fundamental driving force in generating single-wavelength wrinkling. However, surface wrinkling in nature can include more complex patterns such as multiple-length-scale undulations [11,25–27]. To elucidate this feature, we previously proposed a physical model [28,29] that combines membrane bending elasticity and liquid crystal anchoring. A rich variety of multi-scale complex patterns, such as spatial period-doubling and period-tripling are presented for the cases in which the anchoring and bending effects are comparable [28]. In a recent communication [30], we briefly presented a pure higher order anchoring model in the absence of bending elasticity, surprisingly capturing multiple length-scale surface wrinkles. In this previous work, a novel mechanism for the formation of two-scale nano-wrinkling was proposed, which was exclusively based on anchoring energy including quartic harmonics. Here, we present a complete and rigorous new analysis of the multiple-length-scale surface wrinkles based on the pure higher order anchoring model in full detail and approximate the response of the surface structure to chirality and anchoring coefficients based on a linear model. In addition, a fundamental characterization of the capillary vector and capillary pressures required to connect surface geometry and mechanical forces is presented.

The objective of this paper is to identify the key mechanisms that induce and resist the multiple-length-scale surface wrinkling in CLCs based on a pure higher order anchoring model. To develop the anchoring model, we used the generalized shape equation for anisotropic interfaces using the Cahn-Hoffman capillarity [31] and the Rapini-Papoular quartic anchoring energy [32]. The presented model depicts the formation mechanism of two-length scale surface patterns based on the interaction between lower and higher order anchoring modes. The linear approximations of surface curvatures are derived to provide the explicit relations between the anchoring coefficients, helix pitch, and surface profile of the two-length scale wrinkles. These new findings can establish a new strategy for characterizing two-length scale surface wrinkling in biological CLCs, and inspire the design of novel functional surface structures with the potential optical, friction, and thermal applications.

The organization of this paper is as follows. Section 2 presents the geometry and structure of the CLC system. Section 3 presents the governing nemato-capillary shape equation expressing the coupling mechanism between the surface geometry and anisotropic ordering for a CLC free interfaces with a quartic anchoring energy and a pure surface splay-bend deformation. Appendix A presents the details of the derivation of the Cahn-Hoffman capillary vector thermo-dynamics for CLC interfaces. Appendix B describes the capillary shape equation in terms of three capillary pressures. Appendix C represents the shape equation based on the driving and resisting terms. Section 4 analyzes the effect of anchoring coefficients and helix pitch on the surface normal angle and the resultant surface profile. In this section, a general phase diagram of surface profiles in the parametric of anchoring coefficients is presented and the origin of the two scales is revealed through the linear theory. Then, the linear approximations of surface curvatures, assuming small values of anchoring coefficients, are derived to identify the leading mechanism controlling the surface wrinkling. Appendix D proposes the analytical expression for the linear approximation of the surface relief. The surface energy associated with the CLC interface is also analyzed to establish an energy transfer mechanism from anchoring energy of a flat surface into a wrinkled surface. Furthermore, the surface wrinkles are evaluated through analyzing the three capillary pressures, and the pressure–curvature relations are introduced to explore the variation of curvature profile with respect to the capillary pressures. Appendix E represents the derivation of the pressure–curvature relations. Finally, the capillary vectors are formulated to provide a clear physical explanation for the formation of the surface wrinkles. Appendix F formulates the capillary vectors. Section 5 presents the conclusions.

2. Geometry and Structure

Figure 1 depicts the schematics of the CLC structure where ellipsoids indicate fiber orientation on each parallel layer. We assume that the helix axis, **H** is parallel to the surface; other complex structures occurring when the helix axis **H** is distorted are beyond the scope of this paper. The fiber orientation at the interface is defined by the director **n**. The pitch length P_0 is defined as the distance through which the fibers undergo a 2π rotation. For a rectangular (x,y,z) coordinate system, the surface relief that is directed along the x axis can be described by a y(x,z) deviation from the xz plane. The amplitude of the vertical undulation is h(x). As the surface relief is constant in the z direction for a linear texture, the curvature in the z-direction is zero. The unit tangent, **t**, and the unit normal, **k**, to the surface can be expressed with the normal angle, φ: $t(x) = (\sin\varphi(x), -\cos\varphi(x), 0)$, $k(x) = (\cos\varphi(x), \sin\varphi(x), 0)$. L is the given system length in the x direction. The arc-length measure of the undulating surface is "s".

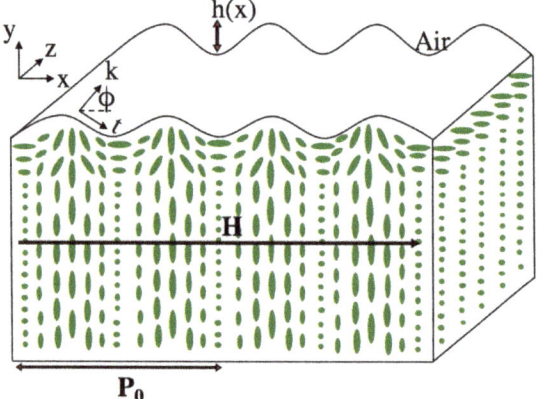

Figure 1. Schematic of a cholesteric liquid crystals (CLC) and surface structures. **H** is the helix unit vector, and P_0 is the pitch. The surface director has an ideal cholesteric twist in the bulk. The helix uncoiling near the surface creates a bend and splay planar (2D) orientation and surface undulations of nanoscale relief h(x) with micron range wavelength $P_0/2$. Adapted from [22].

3. Governing Equations

In this paper, we assume that the multi-length scale surface wrinkles are formed through modulation in surface energy at the anisotropic-air interface of CLCs. The typical capillary shape equations, which are generalized forms of a Laplace equation including the liquid crystal order and gradient density have been comprehensively formulated and previously presented for liquid crystal fibers, membranes, films, and drops [33]. Here, the coupling mechanism between the surface geometry and CLC order are demonstrated through the capillarity shape equation for CLC free interfaces with a pure surface splay-bend deformation.

The formation of surface nanostructures in CLC interfaces is a complex phenomenon involving interfacial tension, surface anchoring energy, and bulk Frank elasticity that requires integrated multi-scale modelling of bulk and surface. However, the analytic solution of the problem with the usual formalism is very complicated. Here, we assume a cholesteric director field in the bulk region, $n_b(x) = (0, \cos\theta, \sin\theta)$, and a splay-bend director field at the interface $n(x) = (\cos\theta, \sin\theta, 0)$ where $\theta = qx, q = 2\pi/P_0$, θ is the director angle, q is the wave vector, and P_0 is the helix pitch.

Based on the generalized Rapini-Papoular equation [24], the interfacial surface energy, γ between a liquid crystal phase and another phase can be described by [32]

$$\gamma = \gamma_0 + \sum_{i=1}^{\infty} \mu_{2i}(\mathbf{n}\cdot\mathbf{k})^{2i} \quad (1)$$

where γ_0 is the isotropic contribution, $\mathbf{n}$ is the director field at the interface, $\mathbf{k}$ is the surface unit normal, and μ_{2i} are the temperature/concentration dependent anchoring coefficients. The preferred orientation that minimizes the anchoring energy (Equation (1)) is known as the easy axis. The actual stationary surface director orientation is the result of a balance between surface anchoring and bulk gradient Frank elasticity [34]. For the cases in which the gradient Frank elasticity is insignificant, the actual stationary and preferred director fields are identical. As shown in ref. [22], for the cholesteric–air interface with quite strong anchoring, the gradient Frank elasticity is negligible in comparison with anchoring in the formation of the surface undulations. It should be noted that here we neglect the Marangoni flow that is likely to be formed due to the orientational-driven surface tension gradients [35–37]. Other effects and processes such as 3D orientation structures, strong nonlinearities, hydrodynamic [38,39], and viscoelastic effects [40–42] discussed elsewhere are beyond the scope of this paper.

The generalized Cahn-Hoffman capillary vector Ξ [43,44], is the fundamental quantity that reflects the anisotropic contribution of CLC in the capillary shape equation. It contains two orthogonal components: normal vector, $\Xi_\perp$ representing the increase in surface energy through dilation (change in area) and tangent vector, $\Xi_\parallel$ representing the change in surface energy through rotation of the unit normal. The derivation details of the Cahn-Hoffman capillary vector thermodynamics for anisotropic interfaces are given in Appendix A [31].

$$\Xi = \Xi_\perp(\mathbf{n},\mathbf{k}) + \Xi_\parallel(\mathbf{n},\mathbf{k}) \tag{2}$$

$$\Xi_\perp = \gamma \mathbf{k}; \quad \Xi_\parallel = \mathbf{I}_s \cdot \frac{\partial \gamma}{\partial \mathbf{k}} \tag{3}$$

Here $\mathbf{I}_s = \mathbf{I} - \mathbf{kk}$ is the 2 × 2 unit surface dyadic, and $\mathbf{I}$ is the identity tensor. The dyadic $(\mathbf{kk})^m$ is similar to $(\mathbf{tt})^m$ due to $(\mathbf{kk})^m = \mathbf{R} \cdot (\mathbf{tt})^m \cdot \mathbf{R}^{-1}$, where $\mathbf{t}$ is the unit tangent and $\mathbf{R}$ is the rotation matrix $\in SO(2)$ satisfying $\mathbf{k} = \mathbf{R} \cdot \mathbf{t}$ (see refs [45] and [46] for details):

$$\mathbf{R} = \begin{bmatrix} 0 & -1 \\ 1 & 0 \end{bmatrix} \tag{4}$$

The following identity holds:
$$\mathbf{nn}:\mathbf{kk} + \mathbf{nn}:\mathbf{tt} = 1 \tag{5}$$

The interfacial static force balance equation at the CLC/air interface is expressed by

$$-\mathbf{k} \cdot (\mathbf{T}^a - \mathbf{T}^b) = \nabla_s \cdot \mathbf{T}_s \tag{6}$$

where $\mathbf{T}^{a/b}$ represent the total stress tensor in the air and the bulk CLC phase, $\nabla_s = \mathbf{I}_s \cdot \nabla$ is the surface gradient operator, and $\mathbf{T}_s$ is the interface stress tensor. The air and the bulk CLC stress tensor, $\mathbf{T}^{a/b}$ are given by

$$\mathbf{T}^a = -p^a \mathbf{I} \text{ and } \mathbf{T}^b = -(p^b - f_g)\mathbf{I} + \mathbf{T}^E \tag{7}$$

where $p^{a/b}$ are the hydrostatic pressures, f_g is the bulk Frank energy density, and $\mathbf{T}^E$ is the Ericksen stress tensor. The bulk Frank energy density for a CLC reads

$$f_g = \tfrac{K_1}{2}(\nabla \cdot \mathbf{n})^2 + \tfrac{K_2}{2}(\mathbf{n} \cdot \nabla \times \mathbf{n} - q)^2 + \tfrac{K_3}{2}(\mathbf{n} \times \nabla \times \mathbf{n})^2 \\ + \tfrac{1}{2}(K_2 + K_4)[\text{tr}(\nabla \mathbf{n})^2 - (\nabla \cdot \mathbf{n})^2] \tag{8}$$

where $\{K_i\}(i=1,2,3)$ are splay, twist, and blend elastic constant, respectively. K_4 is saddle-splay elastic constant. The Ericksen stress tensor, $\mathbf{T}^E$ is given by

$$\mathbf{T}^E = -\frac{\partial f_g}{\partial \nabla \mathbf{n}} \cdot (\nabla \mathbf{n})^T \tag{9}$$

The projection of Equation (6) along direction **k** yields the capillary shape equation:

$$\underbrace{(p^a - p^b) + f_g + \mathbf{kk} : \left[-\frac{\partial f_g}{\partial \nabla \mathbf{n}} \cdot (\nabla \mathbf{n})^T \right]}_{\text{stress jump, SJ}} = \underbrace{(\nabla_s \cdot \mathbf{T}_s) \cdot \mathbf{k}}_{-p_c} \quad (10)$$

where stress jump, SJ, is the total normal stress jump, and p_c is the capillary pressure. Usually we take $p^a - p^b = 0$, and consider the other terms as elastic correction. The interfacial torque balance equation is given by

$$-\mathbf{h} + \mathbf{k} \cdot \frac{\partial f_g}{\partial \nabla \mathbf{n}} = \lambda^s \mathbf{n} \quad (11)$$

where λ^s is the Lagrange multiplier and **h** is the surface molecular field composed by two parts:

$$\mathbf{h} = \underbrace{-\frac{\partial \gamma_{an}}{\partial \mathbf{n}}}_{\mathbf{h}_{an}} \underbrace{-\frac{\partial \gamma_g}{\partial \mathbf{n}} + \nabla_s \cdot \left(\frac{\partial \gamma_g}{\partial \nabla_s \mathbf{n}} \right)}_{\mathbf{h}_g} \quad (12)$$

Here γ_g is the gradient interfacial free-energy density defined by introducing surface gradient energy density vector **g**:

$$\mathbf{g} := (\mathbf{n} \cdot \nabla) \mathbf{n} - \mathbf{n}(\nabla \cdot \mathbf{n}) \text{ and } \gamma_g = \frac{1}{2}(K_2 + K_4) \mathbf{k} \cdot \mathbf{g} \quad (13)$$

By multiplying $(\nabla \mathbf{n})^T$ on both sides of Equation (11), the torque balance equation can be rewritten in a compact form:

$$-\mathbf{h} \cdot (\nabla \mathbf{n})^T + \mathbf{k} \cdot \mathbf{T}^E = 0 \quad (14)$$

Equation (14) gives an alternative path to compute $\mathbf{kk} : \mathbf{T}^E$. The expansion of the term $\mathbf{hk} : (\nabla \mathbf{n})^T$ reads

$$\mathbf{hk} : (\nabla \mathbf{n})^T = \left[-\frac{\partial \gamma_{an}}{\partial \mathbf{n}} - \frac{\partial \gamma_g}{\partial \mathbf{n}} + \nabla_s \cdot \left(\frac{\partial \gamma_g}{\partial \nabla_s \mathbf{n}} \right) \right] \cdot (\nabla \mathbf{n})^T \cdot \mathbf{k} \quad (15)$$

which gives $\mathbf{hk} : (\nabla \mathbf{n})^T = 0$. Thus, only the bulk energy density, f_g, contributes to the elastic correction, which is negligible [22]. For typical cholesteric liquid crystals, the internal length K/γ_0 is in the range 1 nm (an order of magnitude estimation of the elastic constant K and the surface tension γ_0 gives $K \approx 10^{-11}$ J/m and $\gamma_0 \approx 10^{-2}$ J/m^2) [43]. As the ratio of W/γ_0 at the cholesteric–air interface with quite strong anchoring lies in the range (B = W/γ_0 = 0.01), the extrapolation length scale K/W is about $\frac{K}{W} = \frac{\frac{K}{\gamma_0}}{\frac{W}{\gamma_0}} \sim \frac{1\,[nm]}{0.01} \sim 100\,[nm]$. With these values, for a typical CLC with a pitch $P_0 \sim 1.2\,\mu m$, the ratio of extrapolation length scale to pitch is in the order of $\frac{K/W}{P_0} = \frac{20\,[nm]}{1200\,[nm]} = 0.08$. So, the elastic correction contributes 8% to the shape equation, and can be neglected to describe nano-scale surface undulations. As the result, the final shape equation becomes (see Appendix B)

$$p_c = -(\nabla_s . \mathbf{T}_s).\mathbf{k} = \nabla_s.\Xi = \underbrace{\frac{\partial \Xi_\perp}{\partial \mathbf{k}} : (\nabla_s \mathbf{k})}_{\text{dilation pressure}} + \underbrace{\frac{\partial \Xi_\parallel}{\partial \mathbf{k}} : (\nabla_s \mathbf{k})}_{\text{rotation pressure}} + \underbrace{\frac{\partial \Xi_\parallel}{\partial \mathbf{n}} : \nabla_s \mathbf{n}}_{\text{director pressure}} \quad (16)$$

The first two terms contain $\nabla_s \mathbf{k} = -\kappa \mathbf{tt}$, providing information about the surface curvature $\kappa = \frac{d\phi}{ds}$, where ϕ is the normal angle and s is the arc-length. The first term on the right-hand side of Equation (16), which is the usual Laplace pressure, corresponds to the contribution from the normal component of the Cahn-Hoffman capillary vector. The second term which is the anisotropic pressure due to preferred orientation (known as Herring's pressure) corresponds to the contribution from the tangential component of the Cahn-Hoffman capillary vector $\Xi_\parallel$. The last term in Equation (16) represents the additional contribution to the capillary pressure which corresponds to the director curvature due to orientation gradients (see Appendix C). Considering a rectangular coordinate system (x,y,z), where x is the wrinkling direction, and y is the vertical axis, and considering the typical quartic anchoring model [24], $\gamma = \gamma_o + \gamma_a$; $\gamma_a = \mu_2(\mathbf{n} \cdot \mathbf{k})^2 + \mu_4(\mathbf{n} \cdot \mathbf{k})^4$, yields the nonlinear ordinary differential equation (ODE) in terms of normal angle, ϕ:

$$\frac{d\phi}{dx} = \frac{F_{Dr}}{F_{Rs}} = \frac{\partial_s \mathbf{n}}{\sin\phi} \cdot \frac{2[\mu_{2*} + 2\mu_{4*}(\mathbf{n} \cdot \mathbf{k})^2](\mathbf{n} \cdot \mathbf{k})\mathbf{t} + 2[\mu_{2*} + 6\mu_{4*}(\mathbf{n} \cdot \mathbf{k})^2](\mathbf{n} \cdot \mathbf{t})\mathbf{k}}{1 + \mu_{2*}[2(\mathbf{n} \cdot \mathbf{t})^2 - (\mathbf{n} \cdot \mathbf{k})^2] + 3\mu_{4*}(\mathbf{n} \cdot \mathbf{k})^2[4(\mathbf{n} \cdot \mathbf{t})^2 - (\mathbf{n} \cdot \mathbf{k})^2]} \quad (17)$$

Here F_{Dr} denotes as the driving force and F_{Rs} the resistant term. The boundary condition at x = 0 is $\phi|_{x=0} = \frac{\pi}{2}$; μ_{2*} and μ_{4*} are the scaled anchoring coefficients divided by isotropic surface tension γ_0, $\mu_{2*} = \mu_2/\gamma_0$ and $\mu_{4*} = \mu_4/\gamma_0$; and $\widetilde{\phi}(x)$ is the approximation of $\phi(x)$. The generic features of the normal angle and its periodicity are the important outputs of the shape equation. There are three significant system parameters that have influence on the $\phi(x)$: the scaled anchoring coefficients (μ_{2*}, μ_{4*}), and the sign and magnitude of the helix pitch P_0. Thus, the surface profile $h(x)$ is a function of two material properties (μ_{2*}, μ_{4*}) and one structural order parameter (P_0). In the following context, we always assume that helix pitch is constant at $P_0 = 1.2$ μm. Figure 2a depicts the regions with different surface wrinkling in the parametric space of the scaled anchoring coefficients: O_4^+, O_4^-, H_2^+, H_2^-, P_2^+, H_4^+, H_4^-. Here O, H, and P refer to oblique, homeotropic, and planar director anchoring modes, respectively. The reader is directed to reference [30] for a full discussion of these fundamental states. The subscript numbers in O, H, and P indicate the wave number of morphologies in one period, and the superscript sign differentiates the anchoring modes. The transition lines L_1 and L_2 are defined as $L_1 : \mu_{4*} = -\mu_{2*}$ and $L_2 : \mu_{4*} = -\mu_{2*}/2$, and the thermodynamic stability line ($\gamma = 0$) is $S : \mu_{4*} = -1 - \mu_{2*}$. Points A, B, C, and D are chosen as the representative points in P_2^+, O_4^+, H_2^+, and H_4^- regions with $\{\mu_{2*}, \mu_{4*}\}$: A(0.002, 0.001), B(−0.002, 0.0015), C(−0.002, −0.001), and D(0.002, −0.0015). Points A and C, and B and D are related by π rotation symmetry. It should be noted that for the cases in which the quartic anchoring is zero, only single-wavelength sinusoidal profile can be obtained ($\mu_4 = 0$) in the linear regime ($|\mu_{2*}| \ll 1$).

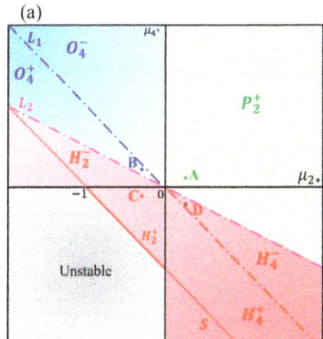 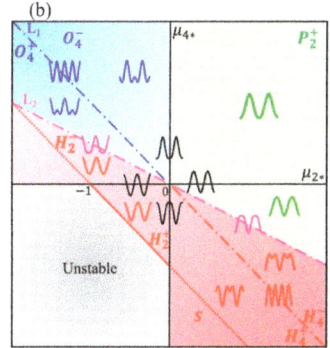

Figure 2. (a) Parametric space in terms of the two anchoring coefficients obtained using Equation (17). Two characteristic lines $L_1 : \mu_{4*} = -\mu_{2*}$ (blue dash-dot) and $L_2 : \mu_{4*} = -\mu_{2*}/2$ (purple dash-dot) indicate wrinkling mode transitions, O_4^+, O_4^-, H_2^+, H_2^-, P_2^+, H_4^+, H_4^-. The subscript numbers indicate how many waves there are within one period. P, O, H represent planar, oblique, homeotropic anchoring, respectively. The thermodynamic line, S, passing point $\mu_{2*} = -1$ illustrates the points where it ends on μ_{2*} and μ_{4*} axes. A (+0.002, +0.001), B (−0.002, +0.0015), C (−0.002, −0.001), and D (+0.002, −0.0015) are four representative points in region P_2^+, O_4^+, H_2^+, and H_4^-, respectively. The region below the dotted S line implies an unstable state because the surface tension is negative. (b) Surface relief profile in the parametric (μ_{2*}, μ_{4*}) space obtained using Equation (17). The anchoring coefficients correspond to all computed curves are less than 0.01.

4. Results and Discussion

4.1. Surface Profile

The surface normal angle, $\phi(x)$ can be directly obtained through solving the governing shape equation, Equation (17). The generic features of the normal angle $\phi(x)$, its magnitude, and its periodicity are the three key outputs of the model. The two significant parameters influencing $\phi(x)$ are the helix pitch P_0, and the scaled anchoring coefficients μ_{2*} and μ_{4*}, which affect the periodicity and the magnitude of $\phi(x)$, correspondingly. Theoretically, μ_{2*} and μ_{4*} give two degrees of freedom to the governing equation. But, for small anchoring coefficients and constant helix pitch, the shape of $\phi(x)$ is only a function of the anchoring ratio, $r = \mu_2/2\mu_4$. The plot of normal angle $\phi(x)$ as a function of the distance "x", corresponding to the points A, B, C, and D, is shown in Figure 3a. As expected, the periodicity equals the half pitch, $P_0/2$, and the amplitude shows a slight deviation, $\phi(x) = \pi/2 + \varepsilon(x)$. Figure 3b shows the effect of helix pitch on the normal angle $\phi(x)$ for the particular point B at three different values of helix pitch P_0, $P_0/2$, and $-P_0/2$. The helix pitch does not influence the amplitude's span of normal angle, but it changes the periodicity of the normal angle. By reducing the helix pitch to half, a more squeezed normal angle profile can be observed. The sign of P_0 reflects the normal angle profile with respect to $\pi/2$. It should be noted that we can estimate the behavior of curvature κ by checking the slope of $\phi(x)$ as $\kappa(x) = \partial_s[\phi(x)] = \phi\prime(x)\sin\phi(x) \approx \phi\prime(x)$.

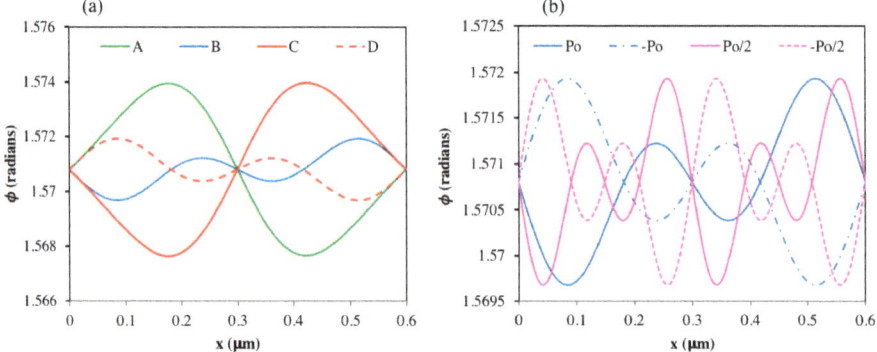

Figure 3. Normal angle profile. (**a**) The normal angle profiles corresponding to the points A, B, C, and D as illustrated in Figure 2a: A (green; mode P_2^+), B (blue, mode O_4^+), C (red, full line, mode H_2^+), and D (red, dashed line, mode H_4^-). (**b**) The normal angle profile for the point B at different helix pitch values of P_0, $-P_0$, $P_0/2$ and $-P_0/2$, where $P_0 = 1.2$ μm.

The surface profile is then obtained from

$$h(s) = -\int_0^s \cot\phi(x)dx \qquad (18)$$

Figure 4a shows typical surface profiles h(x) and corresponding energy profile for the point B and point D. As shown in Figure 4a, increasing P_0 results in both higher periodicity and magnitude. We can clearly see that the surface relief profiles of points B and D exhibit the mirror symmetry, while changing the sign of P_0 result in the same mirror symmetry. These surface undulations can be validated with the two-length-scale surface modulations observed in a sheared CLC cellulosic films [25]. The two different scale periodical gratings include a primary set of bands perpendicular to the shear direction, and a smoother texture characterized by a secondary periodic structure containing "small" bands. It has been shown that the development and periodicities of the small bands are mainly ruled by the CLC characteristics. The chirality of CLC can therefore be mainly responsible for the formation of the secondary bands. The model can be also validated with the two-scale surface pattern of the

Queen of the Night tulip [11], where for this specimen the ratio of amplitudes are $h_2/h_0 = 0.01$, and corresponding wavelength is $\lambda = 1.2$ μm.

Figure 4b shows the scaled energy profile, $\frac{(\gamma_* - 1)}{q}$, in comparison with the surface profile for point B. The scaled energy profile gives the similar plot as the surface relief.

If we denote the parametric vector as $\boldsymbol{\mu}_* = (\mu_{2*}, \mu_{4*})$, then $h(x)$ becomes a function depending on two variables, the vector $\boldsymbol{\mu}_*$ and the helix pitch P_0. Within a linear regime ($|\mu_{2*}| \ll 1, |\mu_{4*}| \ll 1$), the following identities holds true:

$$\text{Geometric Symmetries}: h(\boldsymbol{\mu}_*, P_0) = -h(-\boldsymbol{\mu}_*, P_0) = -h(\boldsymbol{\mu}_*, -P_0) = h(-\boldsymbol{\mu}_*, -P_0) \quad (19a)$$

$$\text{Surface Geometry–Energy Relation}: qh(\boldsymbol{\mu}_*, P_0) = \gamma_*(\boldsymbol{\mu}_*, P_0) - 1 \quad (19b)$$

This identity formulates the symmetric property of surface relief, and its relation to surface energy. Figure 4b is a clear demonstration of symmetry and scaling laws formulated in Equations (19a,b): if we compare B and D we have mirror symmetry and if we plot the anchoring energy of B we would see the same plot as the surface relief: $\underbrace{-h(D, P_0) = h(B, P_0)}_{symmetry}, \underbrace{h(B, P_0) = \frac{\gamma_*(B, P_0) - 1}{q}}_{geometry-energy}$.

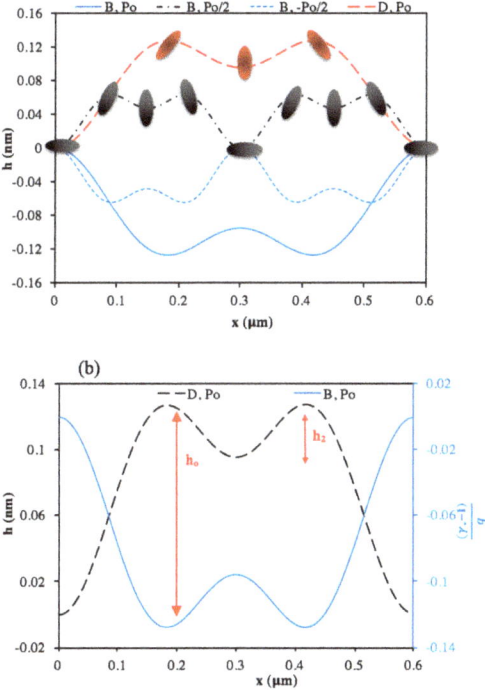

Figure 4. Mirror symmetries observed in surface relief profiles. (a) The surface relief profiles at point B with different helix pitches are given by the two blue curves and the black curve. The red curve gives the surface relief profile at point D. The red and black ellipsoids depict the director orientation for point B with $P_0/2$ and point D with P_0, respectively. These ellipsoids show where the surface extrema occur for planar, homeotropic, and oblique anchoring. (b) The surface profile at point D and scaled energy profile at point B. This figure indicates that there is similarity between surface relief profile and energy profile. The helix pitch is.

Another important parameter that categorizes the shape of surface relief is the ratio between its two wavelengths. The origin of the two scales can be revealed through the linear theory, which gives the signed amplitudes of h_0 and h_2 (the nomenclature is defined in Figure 4b) as a function of anchoring ratio, $r = \mu_2/2\mu_4$:

$$\frac{\tilde{h}_0}{\tilde{h}_2} = \frac{r^2}{(1+r)^2} \tag{20}$$

L_1 and L_2 are defined as the two mode transition lines. Line L_1, which gives a four-wave profile within one period corresponds to the condition $\mu_{4*} = -\mu_{2*}$ ($r = -1/2$, $\tilde{h}_0 = \tilde{h}_2$). Line L_2, which gives a two-wave profile within one period corresponds to the condition $\mu_{4*} = -\mu_{2*}/2$ ($r = -1$, $\tilde{h}_2 \to 0$). In addition, if $\mu_{4*} \to 0$, then $r \to \infty$ such that $\tilde{h}_0 \to \tilde{h}_2$, also gives a two-wave profile.

Figure 2b shows the general phase diagram of h-profiles in the parametric (μ_{2*}, μ_{4*}) plane. As shown in the figure, the transition lines L_1 and L_2 are the critical lines across which surface relief changes its shape. We identify line L_1 as a resonant line with the maximum interaction between quadratic and quartic anchoring effects.

The computations show that h-profile is centrally symmetrical with respect to original point, which can be observed in Figure 2b. As summarized as in Table 1, there are mainly three types of surface wrinkling patterns. It should be noted that there is no difference between O_4^+ and O_4^- as the patterns shown in one region are just a phase shift of the other; the same applies to H_4^+ and H_4^-. However, there is a difference between regions $L_2, \mu_{2*} = 0$ and $H_2^+, H_2^-, P_2^+, \mu_{4*} = 0$ due to the existence of a small plateau shown in the pattern computed along the two lines: L_2 and $\mu_{2*} = 0$. This small plateau corresponds to the discontinuity of two capillary vectors diagram which will be discussed later.

Table 1. Surface wrinkling patterns in different regions of the parametric space $\mu_* = (\mu_{2*}, \mu_{4*})$.

Region	Total Wave Number	h_2/h_0
$O_4^+, O_4^-, H_4^+, H_4^-$	4	$\neq 1$
L_1	4	$=1$
$H_2^+, H_2^-, P_2^+, \mu_{4*} = 0$	2	$=0$
$L_2, \mu_{2*} = 0$	2	$=0$

Results above are considered within one period. Nomenclature: O (oblique), P (planar), and H (homeotropic) refer to the type of anchoring. The L_i's refer to transition lines; see text.

Table 1 summarizes the main four types of surface relief profiles. Region $O_4^+, O_4^-, H_4^+, H_4^-$ and L_1 both give four waves within one period. The difference is that four waves are identical on line L_1. Region $H_2^+, H_2^-, P_2^+, \mu_{4*} = 0$ and $L_2, \mu_{2*} = 0$ both give two waves within one period, so h_2/h_0 is equal to 0. The difference between these two modes is that region $H_2^+, H_2^-, P_2^+, \mu_{4*} = 0$ gives very smooth surface geometry while region $L_2, \mu_{2*} = 0$ gives sharp peaks on the surface profile.

4.2. Surface Curvature

In this subsection we present, discuss, and characterize the surface curvature obtained from direct numerical simulations of the governing equations, and from a new and highly accurate linear model.

The surface behavior is not only affected by the magnitude of the surface relief, but also by the surface curvature. The curvature can be computed directly by two equivalent forms:

$$\kappa = \frac{d\phi}{ds} \text{ or } \kappa = \left[1 + \left(\frac{dh}{dx}\right)^2\right]^{-\frac{3}{2}} \frac{d^2h}{dx^2} \tag{21}$$

The first computing method in Equation (21) is exactly based on the governing Equation (17). Considering that for small values of anchoring coefficients, the resistant term is mainly controlled by isotropic energy γ_0, we obtain the resistant term denoted in Equation (17), $F_{Rs} = 1$. So, the linear approximation of curvature reads

$$\tilde{\kappa}_{\tilde{\phi}} = 2q[\mu_{2*}\cos 2qx + 2\mu_{4*}(3\sin^2 qx \cos^2 qx - \sin^4 qx)] \tag{22}$$

where $\tilde{\kappa}_{\tilde{\phi}}$ denotes the linear approximation of curvature assuming that $\phi = \pi/2$. The analytical expression for the linear approximation of the surface relief is proposed in Appendix D. By assuming $\tilde{\kappa}_{\tilde{h}} = h_{xx}$, we can also obtain another approximation for the surface curvature. It can be easily found that $\tilde{\kappa}_{\tilde{h}} = \tilde{\kappa}_{\tilde{\phi}}$ as we made similar assumptions to approximate the surface curvature based on Equation (21).

A more sophisticated approximation of curvature $\tilde{\kappa}_G$ can be derived without linearizing the governing equation:

$$\tilde{\kappa}_G = q \cdot \frac{2(\mu_{2*} + 6\mu_{4*}\sin^2 qx)\cos^2 qx - 2(\mu_{2*} + 2\mu_{4*}\sin^2 qx)\sin^2 qx}{1 + \mu_{2*}(2\cos^2 qx - \sin^2 qx) + 3\mu_{4*}\sin^2 qx(4\cos^2 qx - \sin^2 qx)} \tag{23}$$

As illustrated in Figure 5, the linear approximation of curvature $\tilde{\kappa}_{\tilde{\phi}}$ obtained by Equation (22) and $\tilde{\kappa}_G$ from Equation (23) provides a very good approximation of curvature. As the curvature $\tilde{\kappa}_{\tilde{\phi}}$ includes the explicit and simple expression, it allows us to mathematically derive more feasible relations to characterize the formation of the surface relief.

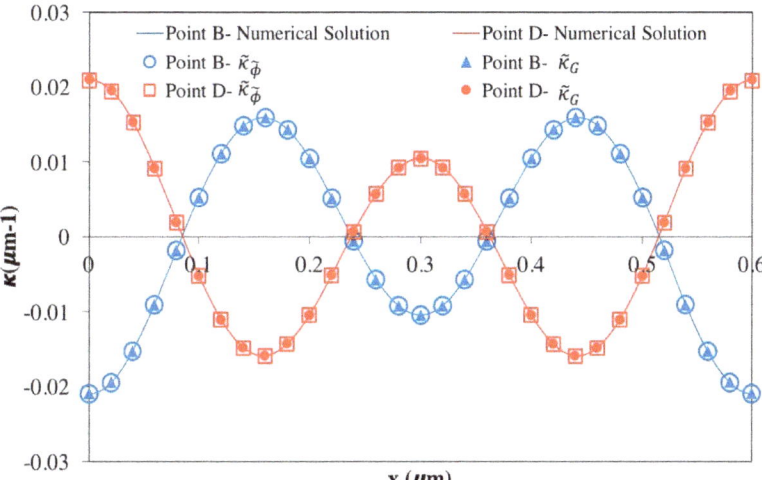

Figure 5. Surface curvature profiles computed numerically and with the two approximation methods: $\tilde{\kappa}_{\tilde{\phi}}$ and $\tilde{\kappa}_G$. Blue and red solid lines are the numerical solutions solved from governing equation for point B and D, respectively. Blue hollow circles and blue filled triangles represent the data points of computed $\tilde{\kappa}_{\tilde{\phi}}$ and $\tilde{\kappa}_G$ at point B, respectively. Red hollow squares and red filled circles represent the data points of computed $\tilde{\kappa}_{\tilde{\phi}}$ and $\tilde{\kappa}_G$ at point D, respectively. As the both approximations $\tilde{\kappa}_{\tilde{\phi}}$ and $\tilde{\kappa}_G$ are identical, the filled circles and triangles are superimposed on hollow squares and circles. The helix pitch is $P_0 = 1.2$ μm.

4.3. Surface Energy

Understanding surface energy behavior is another perspective in realizing the surface profile which helps us to establish an energy transfer mechanism from the anchoring energy of a flat surface into a wrinkled surface. For sufficient small values of the anchoring coefficients, as the normal angle profile $\phi(x)$ is fluctuating around $\pi/2$ with a very small amplitude, an explicit relation between the linearized surface profile and the total surface energy can be estimated based on the linear approximation:

$$\tilde{\gamma}_* - 1 = q\tilde{h} \tag{24}$$

where $(\tilde{\gamma}_* - 1)$ is the scaled anisotropic anchoring energy, and $q\tilde{h}$ is the scaled surface relief. This correlation is detected in Figure 4b where h and $\frac{(\gamma_*-1)}{q}$ are essentially identical for the small anchoring coefficients. This simple expression implies an essential physical phenomenon. The expression, Equation (24) verifies that zero anisotropic surface energy results in a flat surface (h = 0). As the result, based on the expression, the anchoring energy is the driving force contributing to the surface relief, which is in accordance with the previous findings [22]. Moreover, the expression confirms the expected insight that the uppermost surface area contain the highest surface energy.

4.4. Capillary Pressures

As mentioned above, the three main contributions in the capillary pressure are (1) P_{dil}: dilation pressure (Laplace pressure), P_{rot}: rotation pressure (Herrings pressure), P_{dir}: director curvature which is the anisotropic pressure due to the preferred orientation (see Equation (16)). P_{dir} is the driving forces to wrinkle the interface. The explicit expansion of Equation (16) in terms of $(\mathbf{n} \cdot \mathbf{k})$ yields:

$$\begin{aligned}
p_c &= P_{dil} + P_{rot} + P_{dir} \\
P_{dil} &= -\kappa[1 + \mu_{2*}(\mathbf{nn} : \mathbf{kk}) + \mu_{4*}(\mathbf{nn} : \mathbf{kk})^2] \\
P_{rot} &= -\kappa[2\mu_{2*}(\mathbf{nn} : \mathbf{tt} - \mathbf{nn} : \mathbf{kk}) + 4\mu_{4*}(\mathbf{nn} : \mathbf{kk})(3\mathbf{nn} : \mathbf{tt} - \mathbf{nn} : \mathbf{kk})] \\
P_{dir} &= 2[\mu_{2*} + 2\mu_{4*}(\mathbf{nn} : \mathbf{kk})](\mathbf{n} \cdot \mathbf{k})\mathbf{t} \cdot \partial_s \mathbf{n} + 2[\mu_{2*} + 6\mu_{4*}(\mathbf{nn} : \mathbf{kk})](\mathbf{n} \cdot \mathbf{t})\mathbf{k} \cdot \partial_s \mathbf{n}
\end{aligned} \tag{25}$$

As all the pressures are scaled by isotropic tension γ_0, they have the same unit as curvature. It should be noted that based on theory $\dim[P] = \dim[\gamma] \cdot \dim[\partial_s]$.

Figure 6a shows the wrinkling mechanism through the capillary pressures changes along x. The three scaled pressure contributions are plotted as function of "x" for the particular point B. As shown in the figure, the capillary pressures cancel each other out maintaining the summation at zero. The important observation from these pressure profiles is that P_{dil} and P_{dir} are always out-of-phase, while P_{rot} is always negative. These outcomes, $P_{dir} \cdot P_{dil} \leq 0$ and $\text{sgn}(P_{rot}) = -\text{sgn}(P_0)$ can be also interpreted from the linear model. Figure 6a also denotes that P_{rot} is two orders of magnitude smaller than P_{dil} and P_{dir}. This phenomenon confirms that P_{dir} is the formation source of wrinkling, annihilated by inducing area change and area rotation. Another observation from the linear model is that P_{rot} has the similar expression of curvature, $\tilde{\kappa}$. This similarity encourages us that capillary pressures can be also analyzed in the $\kappa - P$ frame. Figure 6b shows the variation of curvature profile with respect to the capillary pressures. We can realize from the figure that in the linear region and for the constant P_0, each capillary pressure only lay on intrinsic curves independent of the anchoring coefficients. The linear approximation gives the intrinsic curves (see Appendix E for the details):

$$\tilde{P}_{dil} = -\tilde{\kappa}, \quad \tilde{P}_{rot} = -\frac{\tilde{\kappa}^2}{q} \quad \text{and} \quad \tilde{P}_{dir} = \tilde{\kappa} + \frac{\tilde{\kappa}^2}{q} \tag{26}$$

The $\kappa - P$ relations approve that helix pitch P_0 is the only parameter affecting the intrinsic curves. Equation (26) implies that the intrinsic curves obtained for $-P_0$ show the central symmetry. Variations in anchoring coefficients do not impose any influence on the intrinsic curves, they only change the arc-length of the intrinsic curves (denoted by $\tilde{l}$). The analytical expression of the arc-length for the intrinsic curves can be obtained by

$$\tilde{l}_{dil} = \sqrt{2}(\tilde{\kappa}_{max} - \tilde{\kappa}_{min}) \tag{27}$$

$$\tilde{l}_{rot} = \left[\kappa \sqrt{\left(\frac{1}{2}\right)^2 + \left(\frac{\kappa}{q}\right)^2} + \frac{q}{4} \ln\left|\frac{\kappa}{q} + \sqrt{\left(\frac{1}{2}\right)^2 + \left(\frac{\kappa}{q}\right)^2}\right| \right]\Bigg|_{\tilde{\kappa}_{min}}^{\tilde{\kappa}_{max}} \quad (28)$$

$$\tilde{l}_{dir} = \frac{q}{4}\left[\left(1+\frac{2\kappa}{q}\right)\sqrt{1+\left(1+\frac{2\kappa}{q}\right)^2} + \ln\left|\left(1+\frac{2\kappa}{q}\right) + \sqrt{1+\left(1+\frac{2\kappa}{q}\right)^2}\right| \right]\Bigg|_{\tilde{\kappa}_{min}}^{\tilde{\kappa}_{max}} \quad (29)$$

where $\tilde{\kappa} \in [\tilde{\kappa}_{min}, \tilde{\kappa}_{max}]$. If we denote $\min_a = -\mu_{4*} - |\mu_{2*} + \mu_{4*}|$, $\max_a = -\mu_{4*} + |\mu_{2*} + \mu_{4*}|$, and $local = -(\mu_{2*} + 2\mu_{4*}) + (\mu_{2*} + 5\mu_{4*})^2/8\mu_{4*}$, then we denote $\min_b = \min\{\min_a, \max_a, local\}$ and $\max_b = \max\{\min_a, \max_a, local\}$, considering the approximation curvature, $\tilde{\kappa}_{\tilde{\phi}}$ (Equation (22)), the interval of $\tilde{\kappa}$ can be found by

$$\tilde{\kappa} \in [2q \cdot \min_a, 2q \cdot \max_a] \text{ if } r \notin [-5/2, 3/2] \quad (30)$$

$$\tilde{\kappa} \in [2q \cdot \min_b, 2q \cdot \max_b] \text{ if } r \in [-5/2, 3/2] \quad (31)$$

These findings denote that the span of curvature is associated with the anchoring coefficients, and ideally exhibits a linear correlation with $1/P_0$. So, we expect that if the helix pitch is increased to $2P_0$ under the same anchoring condition, the span of curvature would reduce to half. Figure 6b illustrates the numerical solutions for director, dilation, and rotation pressures obtained by Equation (25) in comparison with the intrinsic lines defined by Equation (26). We can observe that there are no considerable deviations between the director pressures and the intrinsic lines approximated by the linear model. As shown in Figure 6b, the span of actual curvature is in accordance with the minimum and maximum values of curvature computed by Equations (30) and (31), which confirms that the linear approximation is validated within the linear region (small anchoring coefficients).

In partial summary, in this subsection we have shown (i) the key balancing pressures are the Laplace and director pressures (Figure 6); (ii) quadratic curvature contributions are proportional to the pitch, the curvature–pressure relations follow intrinsic curves (Equation (26)) whose lengths are affected by anchoring, such that lower anchoring (higher anchoring) decreases (increases) their lengths (Equations (27)–(31)).

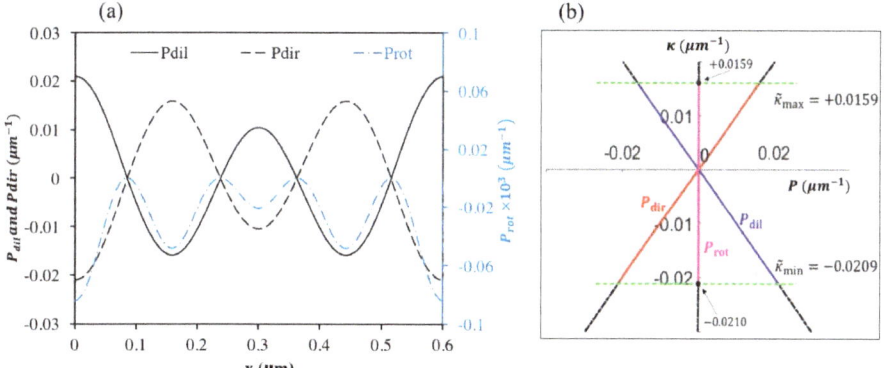

Figure 6. Capillary pressure profile. (**a**) Three components of capillary pressures with respect to x axis for the point B. Black real line, black dash line, and blue dot line represent dilation pressure, director pressure, and rotation pressure, respectively. (**b**) Curvature–Pressure plot at point B. Red, blue, and purple lines represent the numerical solutions to director pressure, dilation pressure, and rotation pressure, respectively. Black dash lines are the intrinsic lines defined by Equation (26). Green dash lines are the span of curvature computed by Equations (30) and (31). Two black points are where the span of numerical solution for curvature ends. The helix pitch is $P_0 = 1.2\mu m$.

4.5. Capillary Vectors

The behavior of the capillary vectors can give another perspective to analyze the surface wrinkling. If we assume that $\mu_{4*} = 0$, then the magnitude of two capillary vectors $\xi_\perp$ and $\xi_\parallel$ naturally satisfy

$$\frac{[\xi_\perp - (\gamma_0 + \frac{1}{2}\mu_2)]^2}{(\frac{1}{2}\mu_2)^2} + \frac{\xi_\parallel^2}{\mu_2^2} = 1 \qquad (32)$$

ξ denotes the magnitude of the capillary vector, Ξ. From this equation, we can read an ellipse with eccentricity $e_{cc} = \sqrt{3}/2$ which is independent of anchoring coefficient μ_2. The two capillary vectors change proportionally; $\xi_\perp$ oscillates around $\gamma_0 + \frac{1}{2}\mu_2$ with an amplitude of $\frac{1}{2}|\mu_2|$, while $\xi_\parallel$ oscillates around zero with amplitude of $|\mu_2|$. This ellipse with invariant shape can provide a clear physical explanation to understand how capillary vectors are formed. Figure 7a illustrates the plots of the ellipse equation for the anchoring coefficient $|\mu_{2*}| = 0.002$. Considering that the CLC surface is differentiable, we can introduce two foci (F_1 and F_2, defined by μ_{2*} in Figure 7b) such that every point P in the vector diagram is restrained by $|PF_1| + |PF_2| = 2|\mu_{2*}|$. When $\mu_{2*} \to 0$, two foci are very close to each other, giving that $|PF_1| \approx |PF_2| = |\mu_{2*}|$. Ellipse becomes a circle with a radius of $|\mu_{2*}|$, which can be considered as a point. From Figure 7a we can also observe that $\xi_\perp$ only reaches its extrema when $\xi_\parallel$ vanishes. This phenomenon corresponds to $\xi_\parallel = \|\mathbf{I}_s \cdot \partial_k \gamma\| = \mathbf{t} \cdot \partial_k \xi_\perp$. However, when $\xi_\parallel$ reaches its extrema, $\xi_\perp$ does not vanish as isotropic surface tension prevents $\xi_\perp$ to be reduced to zero.

The solution to ellipse equation yields

$$\xi_\parallel = \pm\sqrt{\mu_2^2 - 4[\xi_\perp - (\gamma_0 + \frac{1}{2}\mu_2)]^2} \text{ and } \xi_\perp = (\gamma_0 + \frac{1}{2}\mu_2) \pm \frac{1}{2}\sqrt{\mu_2^2 - \xi_\parallel^2} \qquad (33)$$

These are explicit algebraic relations between $\xi_\perp$ and $\xi_\parallel$. Recall that the capillary vectors and the normal angle are related by

$$\phi(x) = \frac{\pi}{2} + \int_0^x \frac{1}{\xi_\perp} \partial_x \xi_\parallel dx \qquad (34)$$

Replacing $\xi_\perp$ with $\xi_\parallel$ from Equation (33), the normal angle can be expressed only in term of $\xi_\parallel$ (see Appendix F):

$$\phi = \frac{\pi}{2} + \phi_\parallel(\xi_\parallel) \text{ where}$$
$$\phi(\xi_\parallel) = 2\arcsin\frac{\xi_\parallel}{\mu_2} - \frac{2(2+\mu_{2*})}{\sqrt{1+\mu_{2*}}}\arctan\left[\sqrt{\frac{1}{1+\mu_{2*}}}\tan\left(\frac{1}{2}\arcsin\frac{\xi_\parallel}{\mu_2}\right)\right] \text{ or} \qquad (35)$$
$$\phi(\xi_\parallel) = -2\arcsin\frac{\xi_\parallel}{\mu_2} - \frac{2(2+\mu_{2*})}{\sqrt{1+\mu_{2*}}}\arctan\left[\sqrt{\frac{1}{1+\mu_{2*}}}\cot\left(\frac{1}{2}\arcsin\frac{\xi_\parallel}{\mu_2}\right)\right]$$

Equation (35) clarifies the source of fluctuation; the perturbation $\phi_\parallel(\xi_\parallel)$, is imposed onto the normal angle profile due to the presence of $\xi_\parallel$, which is fixed by the ellipse equation.

If we assume that $\mu_{2*} = 0$, the magnitude of two capillary vectors $\xi_\perp$ and $\xi_\parallel$ satisfy

$$\frac{\xi_\parallel^2}{\mu_4^2} - 2\left[\frac{4(\xi_\perp - \gamma_0)}{\mu_4}\right]^{\frac{3}{2}} + \frac{16(\xi_\perp - \gamma_0)^2}{\mu_4^2} = 0 \qquad (36)$$

This equation reads a teardrop curve. Figure 7b illustrates the plots of the teardrop equation for the anchoring coefficient $|\mu_{4*}| = 0.002$. The main parameters defining this teardrop curve are given in Figure 7b. Similar to the ellipse curves shown in Figure 7a, the magnitude of μ_{4*} does not change the shape of teardrops, while it controls the size of the teardrop curves. It should be noted that the teardrop curves are not continuous at the original point (Point O shown in Figure 7b). Both the ellipse and teardrop curves show a symmetry by changing the sign of the anchoring coefficients, and shrink to zero as the anchoring coefficients go to zero.

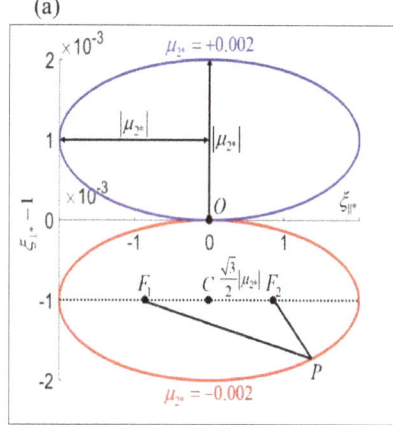 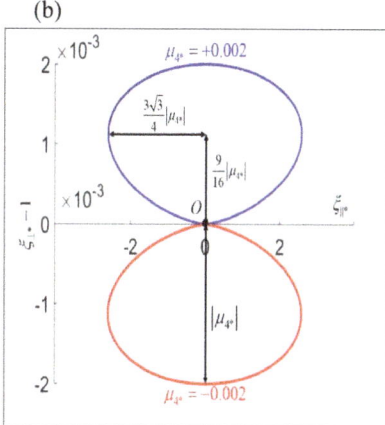

Figure 7. Plots of capillary vector components under two limiting anchoring coefficient values. (a) $\mu_{4*} = 0$ results in an ellipse. (b) $\mu_{2*} = 0$. This results in a teardrop curve. The sign of anchoring coefficient imposes a mirror symmetry. The axes of the loops are determined by the anchoring coefficients.

5. Conclusions

This paper presents a rigorous model based on nonlinear nemato-capillarity shape equation and its linear approximation to describe the main formation mechanism of two-length scale surface wrinkling formed at the CLC/air interface. The role of three capillary pressure contributions (dilation, rotation, and director curvature) on the formation of surface curvature have been elucidated and the effect of the helix pitch and the anchoring coefficients has been characterized. The linear approximation provides a simple model to describe wrinkling behavior with high accuracy and less computation when the two anchoring coefficients are very small. The linear approximation can also serve as the main criteria to classify the type of surface relief. The key mechanism driving surface wrinkling is identified and discussed through the two perspectives: capillary pressures and capillary vectors. Moreover, the surface normal is expressed by the capillary pressures, whose summation must maintain at zero, serving as the constraint to the system. The proposed new model and its linear approximation augment previous models dedicated to understand and mimic complex surface patterns observed at the free surface of synthetic and biological chiral nematic liquid crystals, chiral polymer solutions, surfactant-liquid crystal surfaces and membranes, and in frozen biological plywoods. The present results can inspire design and fabrication of complex surface patterns with the possible potentials in optical, high friction, and thermal applications.

Author Contributions: The manuscript was written through contributions from Z.W and P.R. Review & editing, A.D.R and P.R. All authors have approved the final version of manuscript.

Funding: A part of this work was supported by Le Fonds Quebecois de la Recherche sur la Nature et les Technologies (FRQNT) Postdoctoral Research Scholarship Grant Number 205888.

Acknowledgments: Authors thank the Natural Science and Engineering Research Council of Canada (NSERC) and Le Fonds Quebecois de la Recherche sur la Nature et les Technologies (FRQNT) for financial support of this research.

Conflicts of Interest: The authors declare no conflict of interest.

Appendix A. Cahn-Hoffman Capillary Vector

The purpose of Appendix A is to derive the Cahn-Hoffman capillary vectors that are being used in the main text.

Cahn-Hoffman capillary vector Ξ is defined as the gradient of the scalar field

$$\Xi := \nabla(r\gamma) \tag{A1}$$

$$\mathbf{r} = r\mathbf{k} \tag{A2}$$

where $\mathbf{r}$ is the position vector, and $r = \|\mathbf{r}\|$. $\dim[\Xi] = \dim[\gamma]$. Notice that $d(r\gamma) = \nabla(r\gamma)d\mathbf{r}$:

$$r d\gamma + \gamma dr = \Xi \cdot d(r\mathbf{k}) = r\Xi \cdot d\mathbf{k} + \Xi \cdot \mathbf{k} dr \tag{A3}$$

$\mathbf{t}$ is the unit tangent, $\mathbf{k}$ is the outward unit normal, and $\mathbf{n}$ is director vector (see Figure A1). Equation (A3) yields the two components of capillary vectors (scaled by γ_0):

$$\Xi_{\perp *} = \Xi_{\perp *}\mathbf{k} = \gamma_* \mathbf{k} = \{1 + [\mu_{2*} + \mu_{4*}(\mathbf{nn}:\mathbf{kk})](\mathbf{nn}:\mathbf{kk})\}\mathbf{k} \tag{A4}$$

$$\Xi_{\| *} = \Xi_{\| *}\mathbf{t} = \mathbf{tt} \cdot \frac{\partial \gamma_*}{\partial \mathbf{k}} = [2\mu_{2*} + 4\mu_{4*}(\mathbf{nn}:\mathbf{kk})](\mathbf{nn}:\mathbf{kt})\mathbf{t} \tag{A5}$$

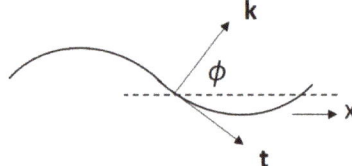

Figure A1. Nomenclature of normal angle and vectors in Cartesian coordinates.

Appendix B. Capillary Pressures

The purpose of Appendix B is to drive the explicit expressions of capillary pressures.

Denote linear operator $\partial_\mathbf{k}(*) = (\mathbf{I} - \mathbf{kk}) \cdot \partial(*)/\partial\mathbf{k}$ for simplicity, and introduce an identity such that:

$$\mathbf{n} \cdot \frac{\partial \mathbf{n}}{\partial s} = 0 \tag{A6}$$

Equation (A6) holds due to $\partial_s(\mathbf{n}^2) = \partial_s(1) = 0$. We can introduce another identity:

$$(\mathbf{I} - \mathbf{nn}) \cdot \frac{\partial \mathbf{M}}{\partial \mathbf{n}} \cdot \frac{\partial \mathbf{n}}{\partial s} = \frac{\partial \mathbf{M}}{\partial \mathbf{n}} \cdot \frac{\partial \mathbf{n}}{\partial s} \tag{A7}$$

Three capillary pressures can be derived by:

$$P_{dir} = \frac{\partial \Xi_{\|*}}{\partial n} : \nabla_s \mathbf{n} = \partial_\mathbf{n} \Xi_{\|*} \cdot \partial_s \mathbf{n} = \partial_\mathbf{n}(\mathbf{t} \cdot \partial_\mathbf{k}\gamma_*) \cdot \partial_s \mathbf{n}$$
$$= 2[\mu_{2*} + 2\mu_{4*}(\mathbf{nn}:\mathbf{kk})](\mathbf{n} \cdot \mathbf{k})\mathbf{t} \cdot \partial_s \mathbf{n} + 2[\mu_{2*} + 6\mu_{4*}(\mathbf{nn}:\mathbf{kk})](\mathbf{n} \cdot \mathbf{t})\mathbf{k} \cdot \partial_s \mathbf{n} \tag{A8}$$

$$P_{dil} = \frac{\partial \Xi_{\perp *}}{\partial \mathbf{k}} : \nabla_s \mathbf{k} = \mathbf{tt} \cdot (\frac{\partial \gamma_*}{\partial \mathbf{k}} \otimes \mathbf{k} + \gamma_* \mathbf{I}) : (-\kappa \mathbf{tt}) = -\kappa \gamma_* \tag{A9}$$

$$P_{rot} = \frac{\partial \Xi_{\|*}}{\partial \mathbf{k}} : \nabla_s \mathbf{k} = \mathbf{t} \cdot \partial_\mathbf{k} \Xi_{\|*} \cdot (-\kappa \mathbf{t}) = \mathbf{t} \cdot \partial_\mathbf{k}[(\mathbf{I} - \mathbf{kk}) \cdot \partial_\mathbf{k}\gamma_*] \cdot (-\kappa \mathbf{t})$$
$$= -\kappa[\mathbf{n} \cdot \partial_\mathbf{k}(\mathbf{n} \cdot \partial_\mathbf{k}\gamma_*) \cdot (\mathbf{nn}:\mathbf{tt}) - (\mathbf{nn}:\mathbf{k}\partial_\mathbf{k}\gamma_*)] \tag{A10}$$
$$= -\kappa[2\mu_{2*}(\mathbf{nn}:\mathbf{tt} - \mathbf{nn}:\mathbf{kk}) + 4\mu_{4*}(\mathbf{nn}:\mathbf{kk})(3\mathbf{nn}:\mathbf{tt} - \mathbf{nn}:\mathbf{kk})]$$

Therefore, the surface curvature is written as:

$$\kappa = \frac{P_{dir}}{\gamma_* + \mathbf{tt} \cdot \partial_\mathbf{k}^2 \gamma_* \cdot \mathbf{t}} = \frac{\partial_\mathbf{n}(\mathbf{t} \cdot \partial_\mathbf{k}\gamma_*) \cdot \partial_s \mathbf{n}}{\gamma_* + \mathbf{n} \cdot \partial_\mathbf{k}(\mathbf{n} \cdot \partial_\mathbf{k}\gamma_*) \cdot (\mathbf{nn}:\mathbf{tt}) - (\mathbf{nn}:\mathbf{k}\partial_\mathbf{k}\gamma_*)} \tag{A11}$$

Equation (A11) is the general curvature equation for $\gamma = \gamma(\mathbf{n} \cdot \mathbf{k})$.

Appendix C. Governing Equation

The purpose of Appendix C is to show how we obtained governing equation and the uniqueness of its solution within the linear region.

A trivial solution to $\nabla \cdot \Xi = 0$ is that Ξ is a constant. Here, we need to consider whether this solution can be true. Inserting boundary condition such that $\phi|_{x=0} = \pi/2$, we get $\Xi = \Xi_\perp = \gamma_0 \hat{\delta}_y$, where $\hat{\delta}_y$ is the unit vector along y axis:

$$\Xi_\perp + \Xi_\parallel = \gamma_0 \hat{\delta}_y \tag{A12}$$

Take the square of (A12) on both side, we obtain:

$$(\gamma \mathbf{k})^2 + (\mathbf{tt} \cdot \partial_\mathbf{k} \gamma)^2 = \gamma_0^2 \tag{A13}$$

This equation does not hold true when both μ_2 and μ_4 are greater than 0. Therefore, Ξ is not a constant. Equation (A11) is an ODE where we can solve for $\kappa(x)$ and subsequently obtain $\phi(x)$:

$$\kappa = \frac{d\phi}{ds} = \frac{d\phi}{dx}\sin\phi = \frac{F_{Dr}}{F_{Rs}} \tag{A14}$$

Here, we introduce two terms, F_{Dr} and F_{Rs} as the driving and resistant terms in the formation of the surface wrinkles that can be defined by:

$$F_{Dr} := 2[\mu_{2*} + 2\mu_{4*}(\mathbf{nn}:\mathbf{kk})](\mathbf{n}\cdot\mathbf{k})\mathbf{t}\cdot\partial_s\mathbf{n} + 2[\mu_{2*} + 6\mu_{4*}(\mathbf{nn}:\mathbf{kk})](\mathbf{n}\cdot\mathbf{t})\mathbf{k}\cdot\partial_s\mathbf{n} \tag{A15}$$

$$F_{Rs} := 1 + 2\mu_{2*}(2\mathbf{nn}:\mathbf{tt} - \mathbf{nn}:\mathbf{kk}) + 3\mu_{4*}(\mathbf{nn}:\mathbf{kk})(4\mathbf{nn}:\mathbf{tt} - \mathbf{nn}:\mathbf{kk}) \tag{A16}$$

Therefore, Equation (A14) becomes the governing equation that is going to be solved to obtain $\phi(x)$. It is easy to find that:

$$F_{Rs,min} - 1 = \min\left\{4\mu_{2*}, -2\mu_{2*} - 3\mu_{4*}, \frac{2}{5}[(3r+4)\mu_{2*} + 6\mu_{4*}]\right\} \tag{A17}$$

where the last term only exists when $-3/2 \leq r \leq 1$. Thus, for very small μ_2 and μ_4, we can conclude that $F_{Rs} > 0$ and it is bounded. So, Equation (A14) has a unique solution.

Appendix D. Linear Theory

The purpose of Appendix D is to show how the linear theory is formed and the application to surface parameters by using linear theory presented in section (3.2).

We consider the first linear theory by simply assuming that $\phi = \pi/2$. Then, governing equation provides a simple expression of surface curvature:

$$\tilde{\kappa}_G = \frac{2(\mu_{2*} + 6\mu_{4*}\sin^2 qx)\cos^2 qx - 2(\mu_{2*} + 2\mu_{4*}\sin^2 qx)\sin^2 qx}{1 + \mu_{2*}(2\cos^2 qx - \sin^2 qx) + 3\mu_{4*}\sin^2 qx(4\cos^2 qx - \sin^2 qx)} \tag{A18}$$

However, we could not obtain an explicit expression for the linearized surface relief $\tilde{h}_P = \int (\int \tilde{\kappa}_P dx) dx$ using this approximated curvature. As we observed that $\Xi_\parallel$ is very small compared to $\Xi_\perp$, the trivial solution can be obtained using an approximation such that $\Xi = \gamma_0 \hat{\delta}_y$.

By using the Monge parametrization, the surface equation can be expressed by:

$$\mathbf{t} = \frac{1}{\sqrt{1+(\nabla_{\|}h)^2}} \begin{bmatrix} 1 \\ \nabla_{\|}h \end{bmatrix} \qquad (A19)$$

Then we multiply by **t** on both side of $\Xi = \gamma_0 \hat{\delta}_y$:

$$\Xi_\perp \cdot \mathbf{t} + \Xi_\| \cdot \mathbf{t} = \gamma_0 \hat{\delta}_y \cdot \mathbf{t} \qquad (A20)$$

$$(2\mu_2 + 4\mu_4 \mathbf{nn} : \mathbf{kk})(\mathbf{nn} : \mathbf{kt}) = \gamma_0 \nabla_{\|} h \qquad (A21)$$

By integrating on both side with respect to x, we obtain:

$$\frac{1}{q}(\mu_2 + \mu_4 \mathbf{nn} : \mathbf{kk})(\mathbf{nn} : \mathbf{kk}) = \gamma_0 h \qquad (A22)$$

Replacing $\phi = \pi/2$, we find the linearized surface relief:

$$\tilde{h} = \frac{1}{q}(\mu_{2*} \sin^2 qx + \mu_{4*} \sin^4 qx) \qquad (A23)$$

Once we get $\tilde{h}$, we can also obtain other surface parameters:

$$\tilde{\phi} = -\mathrm{arccot}(\frac{\partial \tilde{h}}{\partial x}) = -\mathrm{arccot}(\mu_{2*} \sin 2qx + 4\mu_{4*} \sin^3 qx \cos qx) \qquad (A24)$$

Omitting the denominator in Equation (A18), we can obtain an approximation curvature $\tilde{\kappa}_{\tilde{\phi}}$ from the linearized governing equation. We can also approximate surface curvature by $\tilde{h}$:

$$\tilde{\kappa}_{\tilde{\phi}} = \frac{d\tilde{\phi}}{dx} \sin \tilde{\phi} = q[2\mu_{2*} \cos 2qx + 4\mu_{4*}(3\sin^2 qx \cos^2 qx - \sin^4 qx)] \qquad (A25)$$

$$\tilde{\kappa}_{\tilde{h}} = \left[1 + \frac{d\tilde{h}}{dx}\right]^{-\frac{3}{2}} \frac{d^2\tilde{h}}{dx^2} \approx \frac{d^2\tilde{h}}{dx^2} = 2q[\mu_{2*} \cos 2qx + 2\mu_{4*}(3\sin^2 qx \cos^2 qx - \sin^4 qx)] \qquad (A26)$$

Notice that Equations (A25) and (A26) are equivalent.
Equation (A23) also gives a simple explicit form to approximate h:

$$\frac{d\tilde{h}}{dx} = \mu_{2*} \sin 2qx + 4\mu_{4*} \sin^3 qx \cos qx = 0 \qquad (A27)$$

We can also detect the two length scales h_0 and h_2 by finding the extrema of Equation (A27). There are three conditions satisfying (A27): (i) $\sin qx = 0$, (ii) $\cos qx = 0$ and (iii) $\sin^2 qx = -r$. Condition (i) yields that $h = 0$, while the other two conditions yield that $h^* = (\mu_{2*} + \mu_{4*})/q$ and $h^o = -\mu_{2*}^2/4q\mu_{4*}$, respectively. Therefore, we can introduce the following scaling law:

$$\frac{\tilde{h}_0}{\tilde{h}_2} = \frac{h^o}{h^o - h^*} = \frac{-\mu_{2*}^2/4\mu_{4*}}{-\mu_{2*}^2/4\mu_{4*} - \mu_{2*} - \mu_{4*}} = \frac{r^2}{(1+r)^2} \qquad (A28)$$

Equation (A28) indicates that the only parameter r which is defined as the anchoring ratio $r = \mu_2/2\mu_4$ determines the characteristic shape of surface relief.

Appendix E. $P - \kappa$ Relations

The purpose of Appendix E is to demonstrate the relationship between capillary pressures and curvature by using linear theory presented in section (3.3).

Applying linear theory into three capillary pressures, we obtain approximated $P - \kappa$ relations in explicit forms:

$$\tilde{P}_{rot} = -\kappa[2\mu_{2*}(\cos^2 qx - \sin^2 qx) + 4\mu_{4*}(\sin^2 qx)(3\cos^2 qx - \sin^2 qx)] \\ = -\kappa \frac{\tilde{\kappa}}{q} \approx -\frac{\kappa^2}{q} \quad (A29)$$

$$\tilde{P}_{dil} = -\kappa \gamma_* \approx -\kappa \quad (A30)$$

$$\tilde{P}_{dir} = \tilde{\kappa} + \frac{\tilde{\kappa}^2}{q} \quad (A31)$$

Using the linear model, we can find the approximated interval of curvature κ:

$$\tilde{\kappa}/2q = -8\mu_{4*}\cos^4 qx + 2(\mu_{2*} + 5\mu_{4*})\cos^2 qx - \mu_{2*} - 2\mu_{4*} \quad (A32)$$

Since $\cos^2 qx \in [0, 1]$, we need to examine whether local extremum can be achieved or not. Considering function qua(r) as:

$$\text{qua}(r) = -\frac{2(\mu_{2*} + 5\mu_{4*})}{2 \times (-8\mu_{4*})} = \frac{1}{4}r + \frac{5}{8} \quad (A33)$$

This function describes how the x axis of local extrema changes with r.

When local extremum of $\tilde{\kappa}/2q$ is not achieved, there should be qua(r) $\notin [0, 1]$. This gives:

$$r < -\frac{5}{2} \text{ or } r > \frac{3}{2} \quad (A34)$$

In this case, two critical values of $\tilde{\kappa}/2q$ for $\cos^2 qx \in [0, 1]$ would be:

$$\tilde{\kappa}/2q|_{\cos^2 qx=0} = -\mu_{2*} - 2\mu_{4*} \text{ or } \tilde{\kappa}/2q|_{\cos^2 qx=1} = \mu_{2*} \quad (A35)$$

The minimum and maximum of $\tilde{\kappa}/2q$ can be expressed as:

$$\min_a = -\mu_{4*} - |\mu_{2*} + \mu_{4*}| \text{ and } \max_a = -\mu_{4*} + |\mu_{2*} + \mu_{4*}| \quad (A36)$$

If $-5/2 \leq r \leq 3/2$, local extremum becomes:

$$\text{local} = -(\mu_{2*} + 2\mu_{4*}) + \frac{1}{8\mu_{4*}}(\mu_{2*} + 5\mu_{4*})^2 \quad (A37)$$

Let $\min_b = \min\{\min_a, \max_a, \text{local}\}$ and $\max_b = \max\{\min_a, \max_a, \text{local}\}$. Summarizing all the results above, we can find the span of curvature $\tilde{\kappa}$:

$$\tilde{\kappa} \in [2q \cdot \min_a, 2q \cdot \max_a] \text{ if } r \notin [-5/2, 3/2] \quad (A38)$$

$$\tilde{\kappa} \in [2q \cdot \min_b, 2q \cdot \max_b] \text{ if } r \in [-5/2, 3/2] \quad (A39)$$

For example, if we choose point A, we have $r = +1 \in [-5/2, 3/2]$. Then, $\min_a = -0.004$, $\max_a = +0.002$, local $= +0.0021$, and subsequently $\tilde{\kappa} \in [-0.0419, +0.0220]$, where numerical solution gives that $\kappa \in [-0.0422, +0.0221]$. As the linear theory only deviates 0.6% from exact solution, the error can be ignored.

The span in $P - \kappa$ plot can be calculated by arc-length equation:

$$\tilde{l} = \int_{\kappa_{min}}^{\kappa_{max}} \sqrt{1 + \left(\frac{d\tilde{P}}{d\kappa}\right)^2} d\kappa \qquad (A40)$$

Using the fact that:

$$\int \sqrt{a^2 + x^2} \, dx = \frac{x}{2}\sqrt{a^2 + x^2} + \frac{a^2}{2} \ln\left|x + \sqrt{a^2 + x^2}\right| + C \qquad (A41)$$

We can compute the arc-lengths corresponding to the three capillary pressures:

$$\tilde{l}_{dil} = \sqrt{2}(\tilde{\kappa}_{max} - \tilde{\kappa}_{min}) \qquad (A42)$$

$$\begin{aligned}\tilde{l}_{rot} &= \int_{\tilde{\kappa}_{min}}^{\tilde{\kappa}_{max}} \sqrt{1 + \frac{4}{q^2}\kappa^2} d\kappa = 2q \int_{\tilde{\kappa}_{min}}^{\tilde{\kappa}_{max}} \sqrt{\left(\frac{1}{2}\right)^2 + \left(\frac{\kappa}{q}\right)^2} d\left(\frac{\kappa}{q}\right) \\ &= \left[\kappa\sqrt{\left(\frac{1}{2}\right)^2 + \left(\frac{\kappa}{q}\right)^2} + \frac{q}{4}\ln\left|\frac{\kappa}{q} + \sqrt{\left(\frac{1}{2}\right)^2 + \left(\frac{\kappa}{q}\right)^2}\right|\right]_{\tilde{\kappa}_{min}}^{\tilde{\kappa}_{max}-\tilde{\kappa}_{min}}\end{aligned} \qquad (A43)$$

$$\begin{aligned}\tilde{l}_{dir} &= \int_{\tilde{\kappa}_{min}}^{\tilde{\kappa}_{max}} \sqrt{1 + \left(1 + \frac{2\kappa}{q}\right)^2} d\kappa = \frac{q}{2}\int_{\tilde{\kappa}_{min}}^{\tilde{\kappa}_{max}} \sqrt{1 + \left(1 + \frac{2\kappa}{q}\right)^2} d\left(1 + \frac{2\kappa}{q}\right) \\ &= \frac{q}{4}\left[\left(1 + \frac{2\kappa}{q}\right)\sqrt{1 + \left(1 + \frac{2\kappa}{q}\right)^2} + \ln\left|\left(1 + \frac{2\kappa}{q}\right) + \sqrt{1 + \left(1 + \frac{2\kappa}{q}\right)^2}\right|\right]_{\tilde{\kappa}_{min}}^{\tilde{\kappa}_{max}-\tilde{\kappa}_{min}}\end{aligned} \qquad (A44)$$

Using Equation (A42) to Equation (A44), we can compute the arc-length of each $P - \kappa$ curve.

Appendix F. Capillary Vectors

The purpose of Appendix F is to derive the equations related to capillary vectors that are being used in section (3.4).

The following equation holds for capillary vectors:

$$\partial_s \Xi = \partial_s(\Xi_\perp + \Xi_\parallel) = \begin{bmatrix} t & k \end{bmatrix}\left(\begin{bmatrix} \partial_s & -\kappa \\ \kappa & \partial_s \end{bmatrix}\begin{bmatrix} \Xi_\parallel \\ \Xi_\perp \end{bmatrix}\right) \qquad (A45)$$

The projection of capillary vector along **k** direction reduces Equation (A45) to:

$$\nabla_s \cdot \Xi = \partial_s \Xi_\parallel - \kappa \Xi_\perp = 0 \qquad (A46)$$

Replacing $\Xi_\perp$ with $\Xi_\parallel$ from ellipse equation (assuming that $\mu_4 = 0$), we obtain:

$$\frac{d\phi}{ds} = \frac{\partial_s \Xi_\parallel}{\left[(\gamma_0 + \frac{1}{2}\mu_2) \pm \frac{1}{2}\sqrt{\mu_2^2 - \Xi_\parallel^2}\right]} \qquad (A47)$$

Considering that $\mu_2 > 0$, we can use symmetry role for evaluating the cases where $\mu_2 < 0$. Within the linear region, we can integrate on both side of Equation (A47) and obtain:

$$\phi = \frac{\pi}{2} + 2\underbrace{\int_0^{\Xi_\parallel} \frac{1}{\left[(2\gamma_0 + \mu_2) \pm \sqrt{\mu_2^2 - \Xi_\parallel^2}\right]} d\Xi_\parallel}_{\phi(\Xi_\parallel)} \qquad (A48)$$

There are two $\phi(\Xi_\|)$ depending on the positive or the negative sign showed inside the integrand of Equation (A48). The explicit expression of $\phi(\Xi_\|)$ reads:

$$\phi(\Xi_\|) = 2\arcsin\frac{\Xi_\|}{\mu_2} - \frac{2(2+\mu_{2*})}{\sqrt{1+\mu_{2*}}}\arctan\left[\sqrt{\frac{1}{1+\mu_{2*}}}\tan\left(\frac{1}{2}\arcsin\frac{\Xi_\|}{\mu_2}\right)\right] \quad (A49)$$

Assume that $\mu_2 = 0$, $\Xi_\perp$ and $\Xi_\|$ satisfy:

$$\frac{\Xi_\|^2}{\mu_4^2} - 2\left[\frac{4(\Xi_\perp - \gamma_0)}{\mu_4}\right]^{\frac{3}{2}} + \frac{16(\Xi_\perp - \gamma_0)^2}{\mu_4^2} = 0 \quad (A50)$$

We can find that the extrema of $\Xi_\perp - \gamma_0$ are 0 and μ_4. As $\Xi_\perp$ shows symmetry, we only consider the positive branch and find the derivative:

$$\frac{d\Xi_\|}{d\Xi_\perp} = \frac{\mu_4}{2}\left\{2\left[\frac{4(\Xi_\perp - \gamma_0)}{\mu_4}\right]^{\frac{3}{2}} - \frac{16(\Xi_\perp - \gamma_0)^2}{\mu_4^2}\right\}^{-\frac{1}{2}}\left\{\frac{12}{\mu_4}\left[\frac{4(\Xi_\perp - \gamma_0)}{\mu_4}\right]^{\frac{1}{2}} - \frac{32(\Xi_\perp - \gamma_0)}{\mu_4^2}\right\} \quad (A51)$$

The non-trivial extrema of $\Xi_\|$ occur at:

$$\Xi_\perp - \gamma_0 = \frac{9}{16}\mu_4 \text{ and } \Xi_\| = \pm\frac{3\sqrt{3}}{4}\mu_4 \quad (A52)$$

It should be noticed that Equation (A51) is singular at $\Xi_\perp - \gamma_0 = 0$, implying that the plot is not continuous at the point where $\Xi_\perp - \gamma_0 = 0$.

References

1. Vignolini, S.; Moyroud, E.; Glover, B.J.; Steiner, U. Analysing photonic structures in plants. *J. R. Soc. Interface* **2013**, *10*. [CrossRef] [PubMed]
2. Willcox, P.J.; Gido, S.P.; Muller, W.; Kaplan, D.L. Evidence of a cholesteric liquid crystalline phase in natural silk spinning processes. *Macromolecules* **1996**, *29*, 5106–5110. [CrossRef]
3. Sharma, V.; Crne, M.; Park, J.O.; Srinivasarao, M. Structural Origin of Circularly Polarized Iridescence in Jeweled Beetles. *Science* **2009**, *325*, 449–451. [CrossRef] [PubMed]
4. Tan, T.L.; Wong, D.; Lee, P. Iridescence of a shell of mollusk Haliotis Glabra. *Opt. Express* **2004**, *12*, 4847–4854. [CrossRef] [PubMed]
5. Parker, A.R. Discovery of functional iridescence and its coevolution with eyes in the phylogeny of Ostracoda (Crustacea). *Proc. R. Soc. B Biol. Sci.* **1995**, *262*, 349–355.
6. Parker, A.R.; Martini, N. Diffraction Gratings in Caligoid (Crustacea: Copepoda) Ecto-parasites of Large Fishes. *Mater. Today: Proc.* **2014**, *1*, 138–144. [CrossRef]
7. Sharma, V.; Crne, M.; Park, J.O.; Srinivasarao, M. Bouligand Structures Underlie Circularly Polarized Iridescence of Scarab Beetles: A Closer View. *Mater. Today: Proc.* **2014**, *1*, 161–171. [CrossRef]
8. Vukusic, P.; Sambles, J.R.; Lawrence, C.R. Structural colour—Colour mixing in wing scales of a butterfly. *Nature* **2000**, *404*, 457. [CrossRef]
9. Gould, K.S.; Lee, D.W. Physical and ultrastructural basis of blue leaf iridescence in four Malaysian understory plants. *Am. J. Bot.* **1996**, *83*, 45–50. [CrossRef]
10. Graham, R.M.; Lee, D.W.; Norstog, K. Physical and Ultrastructural Basis of Blue Leaf Iridescence in 2 Neotropical Ferns. *Am. J. Bot.* **1993**, *80*, 198–203. [CrossRef]
11. Whitney, H.M.; Kolle, M.; Andrew, P.; Chittka, L.; Steiner, U.; Glover, B.J. Floral Iridescence, Produced by Diffractive Optics, Acts As a Cue for Animal Pollinators. *Science* **2009**, *323*, 130–133. [CrossRef] [PubMed]
12. Urbanski, M.; Reyes, C.G.; Noh, J.; Sharma, A.; Geng, Y.; Jampani, V.S.R.; Lagerwall, J.P.F. Liquid crystals in micron-scale droplets, shells and fibers. *J. Phys. Condens. Matter* **2017**, *29*, 133003. [CrossRef] [PubMed]
13. Mitov, M. Cholesteric liquid crystals in living matter. *Soft Matter* **2017**, *13*, 4176–4209. [CrossRef] [PubMed]

14. Rey, A.D. Liquid crystal models of biological materials and processes. *Soft Matter* **2010**, *6*, 3402–3429. [CrossRef]
15. Bouligan, Y. Twisted Fibrous Arrangements in Biological-Materials and Cholesteric Mesophases. *Tissue Cell* **1972**, *4*, 189–217. [CrossRef]
16. Neville, A.C. *Biology of Fibrous Composites: Development Beyond the Cell Membrane*; Cambridge University Press: New York, NY, USA, 1993; p. vii. 214p.
17. Rey, A.D.; Herrera-Valencia, E.E.; Murugesan, Y.K. Structure and dynamics of biological liquid crystals. *Liq. Cryst.* **2014**, *41*, 430–451. [CrossRef]
18. Rey, A.D.; Herrera-Valencia, E.E. Liquid crystal models of biological materials and silk spinning. *Biopolymers* **2012**, *97*, 374–396. [CrossRef]
19. Murugesan, Y.K.; Pasini, D.; Rey, A.D. Self-assembly Mechanisms in Plant Cell Wall Components. *J. Renew. Mater.* **2015**, *3*, 56–72. [CrossRef]
20. Canejo, J.P.; Monge, N.; Echeverria, C.; Fernandes, S.N.; Godinho, M.H. Cellulosic liquid crystals for films and fibers. *Liq. Cryst. Rev.* **2017**, *5*, 86–110. [CrossRef]
21. Rofouie, P.; Alizadehgiashi, M.; Mundoor, H.; Smalyukh, I.I.; Kumacheva, E. Self-Assembly of Cellulose Nanocrystals into Semi-Spherical Photonic Cholesteric Films. *Adv. Funct. Mater.* **2018**, *28*. [CrossRef]
22. Rofouie, P.; Pasini, D.; Rey, A.D. Nano-scale surface wrinkling in chiral liquid crystals and plant-based plywoods. *Soft Matter* **2015**, *11*, 1127–1139. [CrossRef] [PubMed]
23. Rofouie, P.; Pasini, D.; Rey, A.D. Nanostructured free surfaces in plant-based plywoods driven by chiral capillarity. *Colloids Interface Sci. Commun.* **2014**, *1*, 23–26. [CrossRef]
24. Rofouie, P.; Pasini, D.; Rey, A.D. Tunable nano-wrinkling of chiral surfaces: Structure and diffraction optics. *J. Chem. Phys.* **2015**, *143*, 114701. [CrossRef]
25. Fernandes, S.N.; Geng, Y.; Vignolini, S.; Glover, B.J.; Trindade, A.C.; Canejo, J.P.; Almeida, P.L.; Brogueira, P.; Godinho, M.H. Structural Color and Iridescence in Transparent Sheared Cellulosic Films. *Macromol. Chem. Phys.* **2013**, *214*, 25–32. [CrossRef]
26. Sharon, E.; Roman, B.; Marder, M.; Shin, G.S.; Swinney, H.L. Mechanics: Buckling cascades in free sheets—Wavy leaves may not depend only on their genes to make their edges crinkle. *Nature* **2002**, *419*, 579. [CrossRef]
27. Aharoni, H.; Abraham, Y.; Elbaum, R.; Sharon, E.; Kupferman, R. Emergence of Spontaneous Twist and Curvature in Non-Euclidean Rods: Application to Erodium Plant Cells. *Phys. Rev. Lett.* **2012**, *108*, 238106. [CrossRef]
28. Rofouie, P.; Pasini, D.; Rey, A.D. Multiple-wavelength surface patterns in models of biological chiral liquid crystal membranes. *Soft Matter* **2017**, *13*, 541–545. [CrossRef]
29. Rofouie, P.; Pasini, D.; Rey, A.D. Morphology of elastic nematic liquid crystal membranes. *Soft Matter* **2017**, *13*, 5366–5380. [CrossRef]
30. Rofouie, P.; Wang, Z.; Rey, A.D. Two-wavelength wrinkling patterns in helicoidal plywood surfaces: Imprinting energy landscapes onto geometric landscapes. *Soft Matter* **2018**, *14*, 5180–5185. [CrossRef]
31. Cheong, A.G.; Rey, A.D. Cahn-Hoffman capillarity vector thermodynamics for curved liquid crystal interfaces with applications to fiber instabilities. *J. Chem. Phys.* **2002**, *117*, 5062–5071. [CrossRef]
32. Rapini, A.; Papoular, M. Distorsion d'une lamelle nématique sous champ magnétique conditions d'ancrage aux parois. *J. Phys. Colloq.* **1969**, *30*, C4-54–C4-56. [CrossRef]
33. Rey, A.D. Capillary models for liquid crystal fibers, membranes, films, and drops. *Soft Matter* **2007**, *3*, 1349–1368. [CrossRef]
34. Belyakov, V.A.; Shmeliova, D.V.; Semenov, S.V. Towards the restoration of the liquid crystal surface anchoring potential using Grandgean-Cano wedge. *Mol. Cryst. Liq. Cryst.* **2017**, *657*, 34–45. [CrossRef]
35. Rey, A.D. Nemato-capillarity theory and the orientation-induced Marangoni flow. *Liq. Cryst.* **1999**, *26*, 913–917. [CrossRef]
36. Rey, A.D. Marangoni flow in liquid crystal interfaces. *J. Chem. Phys.* **1999**, *110*, 9769–9770. [CrossRef]
37. Eelkema, R.; Pollard, M.M.; Katsonis, N.; Vicario, J.; Broer, D.J.; Feringa, B.L. Rotational reorganization of doped cholesteric liquid crystalline films. *J. Am. Chem. Soc.* **2006**, *128*, 14397–14407. [CrossRef] [PubMed]
38. Yang, X.G.; Li, J.; Forest, M.G.; Wang, Q. Hydrodynamic Theories for Flows of Active Liquid Crystals and the Generalized Onsager Principle. *Entropy* **2016**, *18*, 202. [CrossRef]

39. Forest, M.G.; Wang, Q.; Zhou, R.H. Kinetic theory and simulations of active polar liquid crystalline polymers. *Soft Matter* **2013**, *9*, 5207–5222. [CrossRef]
40. Brand, H.R.; Pleiner, H.; Svensek, D. Dissipative versus reversible contributions to macroscopic dynamics: The role of time-reversal symmetry and entropy production. *Rheol. Acta* **2018**, *57*, 773–791. [CrossRef]
41. Rey, A.D. Theory of linear viscoelasticity of chiral liquid crystals. *Rheol. Acta* **1996**, *35*, 400–409. [CrossRef]
42. Rey, A.D. Theory of linear viscoelasticity of cholesteric liquid crystals. *J. Rheol.* **2000**, *44*, 855–869. [CrossRef]
43. Hoffman, D.W.; Cahn, J.W. Vector Thermodynamics for Anisotropic Surfaces 1. Fundamentals and Application to Plane Surface Junctions. *Surf. Sci.* **1972**, *31*, 368–388. [CrossRef]
44. Rey, A.D. Mechanical model for anisotropic curved interfaces with applications to surfactant-laden liquid-liquid crystal interfaces. *Langmuir* **2006**, *22*, 219–228. [CrossRef] [PubMed]
45. Fedorov, F.I. Covariant description of the properties of light beam. *J. Appl. Spectrosc.* **1965**, *2*, 344–351. [CrossRef]
46. Sihvola, A.H. Institution of Electrical Engineers. In *Electromagnetic Mixing Formulas and Applications*; Institution of Electrical Engineers: London, UK, 1999; p. xii. 284p.

© 2019 by the authors. Licensee MDPI, Basel, Switzerland. This article is an open access article distributed under the terms and conditions of the Creative Commons Attribution (CC BY) license (http://creativecommons.org/licenses/by/4.0/).

Article

Electro-Thermal Formation of Uniform Lying Helix Alignment in a Cholesteric Liquid Crystal Cell

Chia-Hua Yu, Po-Chang Wu and Wei Lee *

Institute of Imaging and Biomedical Photonics, College of Photonics, National Chiao Tung University, Guiren Dist., Tainan 71150, Taiwan; rickyu@ms69.hinet.net (C.-H.Y.); jackywu@nctu.edu.tw (P.-C.W.)
* Correspondence: wlee@nctu.edu.tw; Tel.: +886-6-303-2121 (ext. 57826); Fax: +886-6-303-2535

Received: 1 March 2019; Accepted: 26 March 2019; Published: 1 April 2019

Abstract: We demonstrated previously that the temperature of a sandwich-type liquid crystal cell with unignorable electrode resistivity could be electrically increased as a result of dielectric heating. In this study, we take advantage of such an electro-thermal effect and report on a unique electric-field approach to the formation of uniform lying helix (ULH) texture in a cholesteric liquid crystal (CLC) cell. The technique entails a hybrid voltage pulse at frequencies f_1 and, subsequently, f_2, which are higher and lower than the onset frequency for the induction of dielectric heating, respectively. When the cell is electrically sustained in the isotropic phase by the voltage pulse of V = 35 V_{rms} at f_1 = 55 kHz or in the homeotropic state with the enhanced ionic effect at V = 30 V_{rms} and f_1 = 55 kHz, our results indicate that switching of the voltage frequency from f_1 to f_2 enables the succeeding formation of well-aligned ULH during either the isotropic-to-CLC phase transition at f_2 = 1 kHz or by the electrohydrodynamic effect at f_2 = 30 Hz. For practical use, the aligning technique proposed for the first time in this study is more applicable than existing alternatives in that the obtained ULH is adoptable to CLCs with positive dielectric anisotropy in a simple cell geometry where complicated surface pretreatment is not required. Moreover, it is electrically switchable to other CLC textures such as Grandjean planar and focal conic states without the need of a temperature controller for the phase transition, the use of ion-rich LC materials, or mechanical shearing for textural transition.

Keywords: dielectric heating; cholesteric liquid crystals; uniform lying helix

1. Introduction

Cholesteric liquid crystals (CLCs) constitute a class of soft photonic crystal, with a self-assembled, helically molecular configuration in one-dimension. Considered in a confined geometry, three main CLC states, namely, Grandjean planar (P), focal conic (FC), and fingerprint (FP), distinctly in molecular orientation and optical properties, have been well distinguished. The molecular helix (helical axis) of a CLC cell in the P state is oriented perpendicular to the substrate normal and it can reflect circularly polarized light with the same handedness as that of the CLC, following Bragg's reflection law. The FC state is found to have randomly oriented molecular helices so that it can scatter unpolarized light due to the mismatch of refractive indices between domain walls. For the FP state where the molecular helix lies in the plane of substrates, unpolarized light passing through it could be either diffracted or transmitted depending on the relation between the wavelength of light and the length of helical pitch [1]. Accordingly, by virtue of unique optical properties and textural switching, CLCs have long been potential candidates for applications in photonics as switchable and memorable optical devices [2–4]. On the other hand, when the helical pitch length is comparable to or shorter than visible light wavelengths, the CLC can also be regarded as a uniaxial birefringent medium analogously to achiral nematic liquid crystals (LCs) with the helical axis being the optical axis. The uniform lying helix (ULH) is a class of this kind of short-pitch CLCs in the FP state with the helical helix oriented in the substrate plane along a preferred direction. In contrast to the aforementioned CLC textures, the ULH is attractive for its superior electro-optic responses, enabling the modulations in phase retardation and intensity of polarized light under crossed polarizers by the processes of helix-unwinding and in-plane switching of optic axis via the voltage-induced dielectric and flexoelectric effects, respectively [5]. Particularly, the flexoelectro-optic switching of ULH structures with a wide viewing angle and short response time (<1 ms) have been proven by extensive studies, including those using bimesogenic LCs [6–8] and others concerning doping bent-core dimers into rod-like CLCs [9–11]. Owing to the luring uniqueness, the ULH structure has received a lot of interest, extending the application domain of CLCs to next-generation displays, such as ultra-high resolution TVs, virtual reality and augmented reality head-mounted displays, glasses-free 3D displays, and field-sequential color displays, whose images need to be displayed with high definition and high refresh rate.

The most severe problem impeding practical uses of the ULH is the difficulty in obtaining a defect-free alignment in a simple cell geometry. This is not surprising because a CLC in the cell with either planar or homeotropic alignment surfaces is favorably stabilized in the P or FC state to minimize the free energy. For this purpose, technologies based on electric-field and surface-alignment pretreatments to induce the ULH alignment have successively been exploited with a varying degree of limitations. The earliest and the most commonly adopted electric-field method was reported by Lee and Patel, in which a temperature-cooling process is essential for the CLC cell to go through the isotropic-to-CLC phase transition when an AC voltage is applied [12]. The optic axis of the obtained ULH is formed with a deviation angle from the rubbing direction. On the basis of this method, Salter et al. reported that the resulting ULH in a planar-aligned cell with 90°-twisted rubbing is more uniform than those of hybrid-aligned cells and of the planar-aligned cells with parallel (0° or 180°) rubbing [13]. Without undergoing the temperature-cooling process, voltage-generated ULH alignment has also been obtained in planar-aligned CLC cells by providing an extra mechanical force to induce shear flow in the meantime [14–16], using ion-rich positive nematic host to permit electrohydrodynamic flow by low-frequency voltages [17,18], controlling voltage conditions precisely to induce the nematic-to-CLC structural transition [19], optimizing the pretilt angle as well as the anchoring energy [20], or by designing a tri-electrode configuration [21,22]. For the surface-treatment method, the concept stems from the modification in surface morphology of alignment layers, enabling periodic anchoring for realizing the ULH structure spontaneously. This prerequisite has been implemented by using cholesteric alignment layers with the same pitch as that of injected material [23], coating weak homeotropic alignment films on scratched substrates [24], generating interdigital planar and homeotropic channels on substrates [25], and creating grooves by means of two-photon excitation laser lithography [26], mold-templating [27], and by laser writing [28]. In some cases of the above, the isotropic-to-CLC temperature cooling process with and without the assistance of the electric field was required to ensure the uniformity of ULH. However, the manufacturing processes of ULH by surface-treatment methods would become too complicated, especially for mass production.

Recently, we have demonstrated that the dielectric heating effect can be induced in a non-dual-frequency CLC cell by high-frequency voltages (e.g., negative chiral nematic in [29]). As confirmed experimentally and theoretically, this is caused by the pseudo-dielectric relaxation from the non-ideal cell geometry with finite conductivity in indium-tin-oxide (ITO) electrodes rather than the intrinsic molecular rotation. Based on this electro-thermal effect, we propose, in this work, to form ULH structures by applying a hybrid pulse with a fixed amplitude and frequencies f_1 and f_2 ($<f_1$) across a CLC cell thickness. The pulse at f_1 is responsible for elevating the cell temperature. Switching the frequency from f_1 to f_2 enables the formation of ULH alignment during the phase transition or via the electrohydrodynamic instability. In the following, we first clarify the mechanism of the voltage-induced textural and phase transition of the CLC by dielectric heating. The processes of forming ULH alignment by the treatment of designated hybrid pulse are then manifested.

2. Experimental

The CLC used in this study was a mixture composed of 70 wt% of a positive nematic LC (E44, Daily Polymer Co., Kaohsiung, Taiwan) and 30 wt% of the left-handedness chiral additive S811 (HCCH). Material properties of E44 give the refractive indices $n_{||}$ = 1.79 and $n_\perp$ = 1.53 (measured at the wavelength of 589 nm and temperature of 20 °C), the twist elastic constant K_{22} of 13 pN, and dielectric anisotropy $\Delta\varepsilon$ = 14.38 as determined by the difference between the parallel and perpendicular components of dielectric constant of $\varepsilon_{||}$ = 19.42 and $\varepsilon_\perp$ = 5.04, respectively (at the frequency of 1 kHz and temperature of 20 °C). The CLC was put on a hot stage, heated to the isotropic phase, stirred for two hours, and finally injected by capillary action into an empty cell with a planar surface alignment and 90°-twisted rubbing [19]. Cell parameters, including the electrode area $A = 10 \times 10$ mm^2, cell gap $d = 4.8 \pm 0.5$ μm, and the ITO electrode resistivity R_{ITO} ~ 310 Ω, were identical to those reported in [29] so that the dielectric heating effect could be expectedly induced by optimized voltage conditions. The clearing point T_c (i.e., the transition temperature between the isotropic and CLC phases) of the CLC cell, as determined by temperature-dependent dielectric spectroscopy together with textural observations, was 57.0 °C and the CLC phase could be preserved at least down to 0 °C. Dielectric spectra of the CLC cell at designated temperatures were acquired with an LCR meter (Agilent E4980A) and a temperature controller (Linkam T95-PE). The adjustable amplitude and frequency ranges of the sinusoidal probe voltage provided by the LCR meter were 0.05 V − 20 V and 20 Hz−2 MHz, respectively. The CLC cell was driven by square-wave voltage supplied from an arbitrary function generator (Tektronix AFG-3022B) in conjunction with an amplifier (TREK Model 603). The types of CLC textures and their uniformity were preliminarily examined by textural observation using a polarizing optical microscope (Olympus BX51) in the transmission mode. The optical transparency or voltage-dependent transmission of CLC textures was quantitatively inspected by setting the cell between a He−Ne laser source with an emission wavelength of 632.8 nm and a photodetector without any polarizer. A non-contact IR thermometer (IR camera FLIR ThermaCam® P25) was employed to monitor the variation in cell temperature as a function of the applied voltage.

3. Results and Discussion

3.1. Frequency-Modulated Textural and Phase Transitions

Figure 1a shows the voltage-dependent transmission (V–T%) curves of the CLC cell at a fixed temperature of T = 25 °C as measured without any polarizer. Here, the temperature was precisely controlled by situating the cell in a temperature controller. In the case of the frequency at f = 1 kHz, the resulting V–T% curve illustrates the change of CLC texture from the initial P state to the FC state and finally to the homeotropic (H) state with increasing voltage. This follows the general case for the voltage-induced textural transition of a CLC cell with positive dielectric anisotropy according to the dielectric coupling between LC molecules and electric fields. Accordingly, the V–T% curves at given frequencies in Figure 1a can be divided into five voltage regions, separated by colors, corresponding to specific CLC states and textural transitions. Because the wavelength of the light source (i.e., He–Ne laser) at 632.8 nm is outside the (Bragg) reflection bandgap of the CLC, the transmittance (~80%) of the cell in P state (~80% in region 1, V < 6 V$_{rms}$) is comparable to that of the voltage-sustained H state (~82% in region 5, V > 38 V$_{rms}$). The voltage range (region 3, 6 V$_{rms}$ < V < 38 V$_{rms}$) resulting in a much-lowered transmittance (T% < 8%) corresponds to the cell's FC state that effectively scatters visible light (at the wavelength of 632.8 nm) due to its randomly oriented and broken helices in the bulk. The voltage region 2 (4), showing decreased (increased) transmittance from 80% (5%) to 5% (82%) with ascending voltage between 6 V$_{rms}$ (28 V$_{rms}$) and 28 V$_{rms}$ (38 V$_{rms}$), is thus attributable to the P-to-FC (FC-to-H) textural transition. Besides, in a special case where the frequency of the applied voltage is low enough to bring about electrohydrodynamic instability (EHDI), the ULH state can be obtained in a given voltage region. This phenomenon has been evidenced in some of the early works [17,18,30]. However, when the frequency varies from 1 kHz to 30 Hz in this study, the V–T%

curve remains nearly unchanged as shown in Figure 1a. This implies that the ionic effect in our used CLC cell is weak and the frequency of the voltage at 30 Hz is still insufficient to onset the EHDI; thus, the optical signal corresponding to the ULH state is absent in Figure 1a.

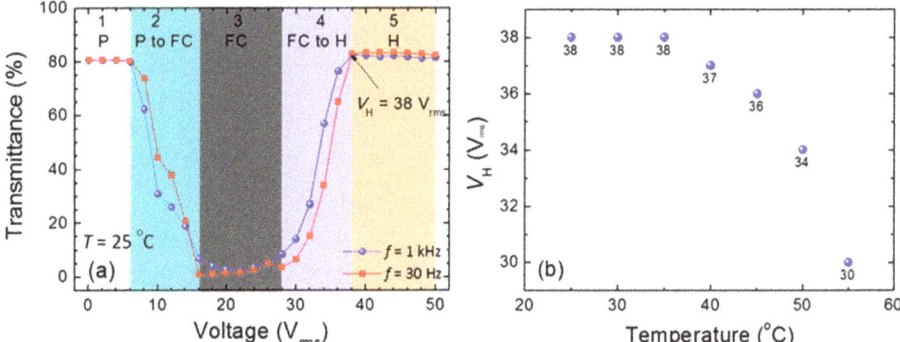

Figure 1. (a) Voltage dependence of the transmission of the CLC cell under AC voltage applied at $T = 25\ °C$ and (b) the change of the critical voltage (V_H) for helix unwinding as a function of temperature. Regions 1, 3, and 5 in Figure 1a are voltage ranges for sustaining the cell in the P, FC, and H states, respectively. Regions 2 and 4 reveal V–T% behaviors, dominated by the P to FC and FC to H textural transitions, respectively.

The voltage V_H required for unwinding the CLC helix as well as sustaining the cell in the H state can be expressed as

$$V_H = \frac{\pi^2 d}{P}\sqrt{\frac{K_{22}}{\varepsilon_0 \Delta \varepsilon}} \tag{1}$$

where P stands for the helical pitch length and ε_0 (= 8.854×10^{-12} F·m^{-1}) represents the permittivity in a vacuum. One can see from Figure 1b that the magnitude of V_H decreases from 38 V_{rms} to 30 V_{rms} as the temperature increases from 25 °C to 55 °C. In most of thermotropic calamitic or rod-like LCs, the orientational order parameter (S) decreases with increasing temperature. Since the twist elastic constant K_{22} and the dielectric anisotropy $\Delta \varepsilon$ are proportional to S^2 and S, respectively, the change in V_H as a function of temperature is primarily attributable to the substantial variation in temperature-dependent K_{22}.

Furthermore, Figure 2 depicts the applied (voltage) frequency dependence of the steady-state cell temperature and transmission (Figure 2a) as well as optical textures of the CLC cell subjected to a 35-V_{rms} voltage at three distinct frequencies (Figure 2b–d). Here the measurements were performed without using a temperature controller. The cell temperature in the field-off state was ca. 24.4 °C, namely, the room temperature. Each cell-temperature data point was acquired by holding the given voltage applied to the cell for 10 min to ensure that the cell temperature reached the steady state and became time-independent. It was tested preliminarily that the onset frequency f_{th} for the induction of dielectric heating by the applied voltage at $V = 35\ V_{rms}$ is approximately 10 kHz. As revealed in Figure 2a, the temperature of the cell grows with ascending frequency but this trend becomes mitigated as the frequency is higher than 55 kHz. This is in good agreement with our previous work in which the temperature increase, arising from significant dielectric heating from the pseudo-dielectric relaxation, is a function of the applied frequency [29,31]. In the case of $f = 20$ kHz, the temperature is elevated by 3.6 °C (i.e., $T = 28,0\ °C$ at $f = 20$ kHz); thus, the texture (Figure 2b) under applied voltage $V = 35\ V_{rms}$ (lower than $V_H = 38\ V_{rms}$ at $T = 30.0\ °C$), corresponding to region 4 in Figure 1a, is preserved in an intermediate state between FC and H state with low transmission of ~27.5%. As the frequency is promoted from 20 kHz to 45 kHz, the temperature increases monotonically from 28.0 °C to 46.4 °C. In the meantime, the driving voltage (i.e., V = 35 V_{rms}) approaches V_H = 36 V_{rms} at T = 45 °C. In consequence, the sharp increase in transmittance to ~78% with increasing frequency in the neighborhood of 45 kHz, as shown in Figure 2a, is undoubtedly contributed by the process of the FC-to-H phase transition (Figure 2c). Once the frequency goes beyond 55 kHz such that the temperature showing T = 58.0 °C becomes higher than the clearing temperature (T_c = 57.0 °C), the phase transition from the CLC phase to the isotropic phase will be induced. Although the optical textures under crossed polarizers and transmittances without a polarizer in both the isotropic phase and the H state are comparable or virtually identical, the voltage-induced isotropic phase can be confirmed by comparing the optical textures between the electrode and non-electrode areas. As evidenced in Figure 2d, the texture of the cell in the non-electrode regions around the electrode area gradually changes from the CLC to the isotropic phase owing to the diffusion of heat generated from the electrode. This observation suggests that the textural and phase transitions in the CLC cell can alternatively be induced by modulating the frequency of the applied voltage to control the cell temperature via dielectric heating [29]. From the results above, one can form a well-aligned and switchable ULH texture by treating a CLC cell with a specific AC voltage.

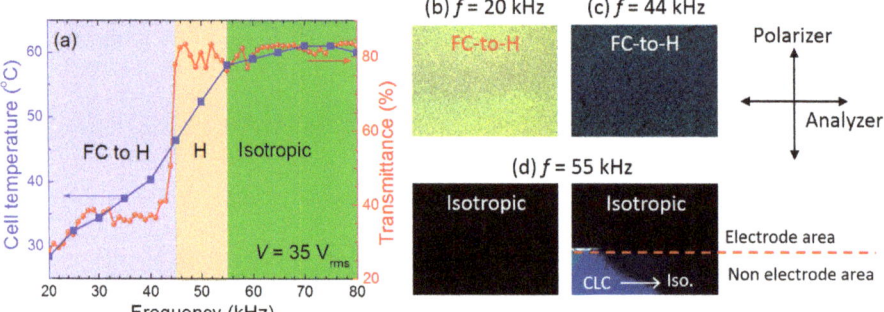

Figure 2. (a) Frequency-dependent cell temperature and transmission of the CLC cell driven by a square-voltage at V = 35 V_{rms} and the corresponding optical textures at (b) f = 20 kHz, (c) f = 44 kHz, and (d) f = 55 kHz. For (b–d), the transmission axes of the analyzer (A) and the polarizer (P) are crossed to each other.

Figure 3 delineates the idea of the proposed electric-field approach to the formation of well-aligned ULH and the scheme for the textural switching. Here, an applied voltage signal, consisting of two pulse components with one at f_1 followed instantly by the other at a lowered frequency f_2, is designed for carrying out the switching of the cell from either the P or FC state to the ULH state. The magnitudes of f_1 and f_2 designated satisfy the conditions of $f_1 > f_{th}$ and $f_2 < f_{th}$, where f_{th} is the onset frequency for the induction of the dielectric heating in the cell. Such a pulse functions like a heat generator at f_1, enabling the temperature increase of the cell to a given value; it also serves as an agitator at f_2 to provoke phase and textural transitions. As such, there are two ways, alternative to ideas reported in [12,17], to form the ULH alignment by optimizing voltage conditions of the hybrid pulse. Referring to Figure 2a, the first way is to first sustain the cell in the isotropic phase (at an arbitrary temperature beyond the clearing point) by setting the amplitude and frequency of the leading portion of the pulse to be V = 35 V_{rms} and f_1 = 55 kHz. When the pulse frequency is switched from f_1 to f_2 whereas the voltage magnitude remains unchanged, the dielectric heating behavior disappears, and the ULH results at f_2 during the isotropic-to-CLC phase transition. In comparison with the method reported in [12], using the designated hybrid pulse to form ULH structure would be more feasible because an additional heating stage is not required to monitor the temperature variation for the phase transition. Our method further permits the ULH to be electrically switchable by following the general scheme for switching CLC textures. As illustrated in Figure 3, the obtained ULH can be readily switched back to the P state by triggering the cell with a single pulse of $V = V_H$ at $f = f_2$, and to the FC state by a hybrid pulse at f_2 with amplitudes of V_H for the leading component and $V < V_H$ for the succeeding component. The second way is to generate ULH via EHDI by elevating the temperature to promote the ionic effect rather than requiring the LC material to be ion-rich [17,18]. In this case, the cell is held in the H state at a temperature close to T_c by the first portion of the pulse at $V = 30$ V$_{rms}$ and $f_1 = 55$ kHz. It has been demonstrated that the ionic effect in an LC cell is enhanced at a higher temperature thanks to the reduction in molecular ordering and increase in the activation energy of mobile ions [32]. Therefore, it is expected to obtain ULH by optimizing f_2 to meet the frequency responsible for inducing the EHDI.

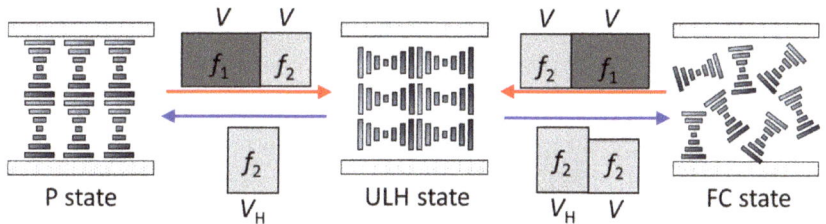

Figure 3. Schematic of the generation of the ULH state and the scheme for textural switching after voltage treatment. Note that $f_1 > f_2$.

3.2. Formation of ULH Alignment after the Treatment of a Designated Hybrid Pulse

Following the switching scheme illustrated in Figure 3, Figure 4 shows the processes of the generation of ULH alignment in the CLC cell driven by a hybrid voltage. The total duration of the applied pulse is 810 s. In the case of the ULH formation through the isotropic-to-CLC phase transition, $V = 35$ V$_{rms}$ and $f_1 = 55$ kHz are fixed. As shown in Figure 4a, the cell is brought to the isotropic phase by the pulse at f_1 for 600 s. Once the pulse is switched from f_1 to $f_2 = 30$ Hz or 1 kHz, the isotropic-to-CLC phase transition occurs within 30 s. Meanwhile, the ULH starts to appear and is completely formed throughout the cell at time $t = 810$ s. The optic axis of this ULH alignment is deviated by an angle of ~45° from the rubbing direction because of the use of a 90°-twisted planar-aligned cell [13,19]. As a result, by rotating the optic axis of ULH parallel to the transmission axis of either of crossed polarizers, excellent dark appearance can be secured, indicating the uniform ULH alignment. Note that the ULH generated is a metastable state which can be retained for several hours at room temperature after voltage removal. The stability of the ULH can further be promoted by optimizing the cell-gap-to-pitch ratio [17] or by performing polymer stabilization in addition. When V is reduced from 35 V$_{rms}$ to 30 V$_{rms}$, the temperature of the cell driven by the pulse at f_1 for 600 s, as determined by a thermograph camera, is about 55.2 °C, which is lower than $T_c = 57.0$ °C. The dark state of the cell under crossed polarizers thus corresponds to the H state. Because $V = 30$ V$_{rms}$ is very close to $V_H = 30$ V$_{rms}$ at $T = 55$ °C (Figure 1b), the textural transition is induced quickly after switching the pulse frequency to $f_2 = 30$ Hz for 10 s (i.e., $t = 610$ s in Figure 4b), allowing the ULH to be obtained within the next 20 s. However, if f_2 is changed from 30 Hz to 1 kHz, the FC state instead of the ULH state is generated (Figure 4b). This implies that the ULH formed by the 30 Hz voltage pulse might be a result of the electrohydrodynamic effect (to be discussed later).

The results mentioned above are further substantiated by measuring the time-dependent transmission curves of the CLC cell within the time of the voltage pulse at f_2. As shown in Figure 5a, except for the case where the cell exhibits relatively low transmittance (~26%) due to the formation of FC state by $V = 30$ V$_{rms}$ at $f_2 = 1$ kHz, the transmittances of the other three textures with ULH alignment in the entire measured time are higher than 70%, which are only about 10% lower than that of the H state. The wavy variation in specific temporal regions in these three curves might be attributable to the phase or the textural transition and to the variation in refractive indices with temperature over time. Figure 5b displays the time-dependent temperature and temperature increment of the cell subjected to

30- and 35-V$_{rms}$ hybrid pulses. Driving the cell at f_1 = 55 kHz causes the cell temperature to increase, reaching saturation at T = 58.0 °C for V = 35 V$_{rms}$ and T = 55.2 °C for V = 30 V$_{rms}$ at $t \sim$ 200 s; thus, holding the cell in the isotropic phase and the H state in the CLC phase, respectively. At the time when the pulse is changed from 55 kHz to f_2 = 30 Hz, the temperature drops with ascending time and returns to the initial value (i.e., T = 24.4 °C) within 300 s ($t \sim$ 900 s). The result revealed in Figure 5b also indicates that the isotropic-to-CLC phase transition at V = 35 V$_{rms}$ and the H-to-CLC textural transition can be induced within 30 s after the frequency is switched from f_1 to f_2.

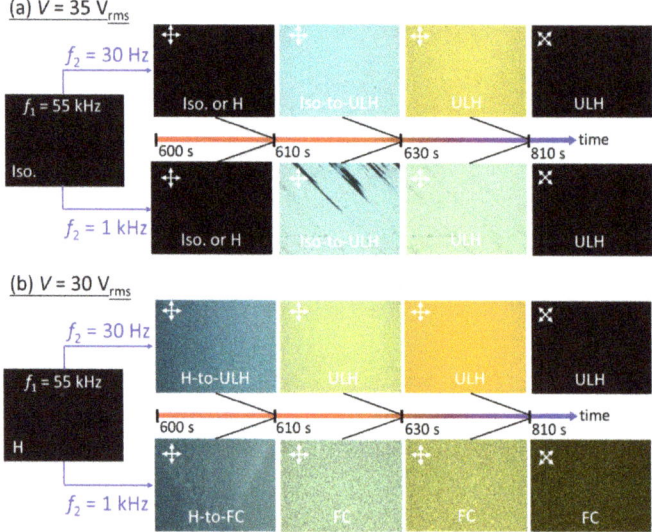

Figure 4. Processes of the formation of ULH alignment in the CLC cell during the application of separate hybrid pulses of (**a**) V = 35 V$_{rms}$ and (**b**) V = 30 V$_{rms}$ at f_1 = 55 kHz and f_2 = 30 Hz or 1 kHz. The duration at f_1 lasts for 600 s.

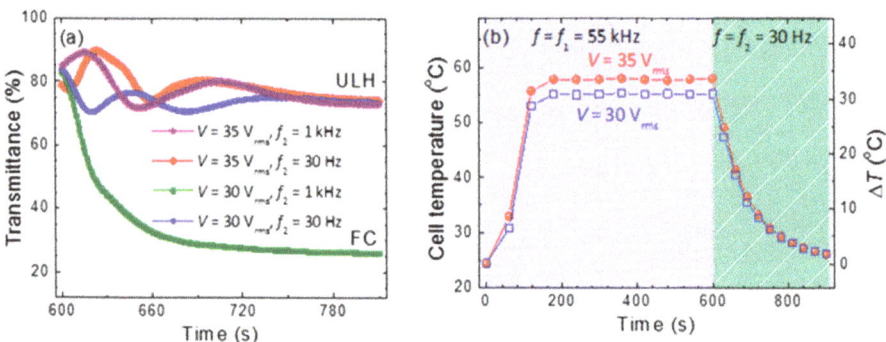

Figure 5. (**a**) Time-dependent transmission curves of the cell after the switching of the frequency of the hybrid pulse from f_1 (= 55 kHz) to f_2 and (**b**) time-dependent temperature and temperature variation of the cell during the application of hybrid pulses at V = 35 V$_{rms}$ and V = 30 V$_{rms}$. The duration of the pulse at f_1 is 600 s.

Based on the results mentioned above, it should be emphasized here that the ambient temperature would not be an issue affecting the generation of ULH alignment by the hybrid voltage pulse. It has been proven that the strength of temperature elevation by dielectric heating can be predominated by voltage conditions [29], as shown in Figure 2a for example. Therefore, no matter how the ambient temperature is, the cell temperature can be increased to an appropriate value (e.g., a temperature higher than and near the clearing point T_c of the CLC material for inducing ULH alignment via phase and textural transitions, respectively) by optimizing the frequency and the amplitude of the leading component of the hybrid voltage pulse (i.e., the pulse at f_1). Moreover, according to the results of Figures 4 and 5, when the frequency of the voltage pulse is switched from f_1 to f_2, the ULH alignment will be formed within the temperature range near but below T_c, which is independent of the ambient temperature. The total elapsed time to form ULH alignment by the proposed method can be cut down at the expense of increasing the magnitude V at f_1 (to shorten the heating time). However, such a waveform design in the operation pulse may be considered complicated in that V at f_2 needs to be low enough to permit the induction of phase or textural transition.

To explain that the ULH, formed by the hybrid pulse of $V = 30$ V_{rms} at $f_1 = 55$ kHz and $f_2 = 30$ Hz, is enabled and governed by EHDI, Figure 6a,b show the real (ε')- and imaginary (ε'')-part dielectric and the loss-tangent (tanδ) spectra of the CLC cell in the P state at $T = 25$ °C and 50 °C, respectively. The probe voltage for dielectric measurements is as low as 0.5 V_{rms}, which is unable to give rise to the Fréedericksz transition and, in turn, the molecular reorientation. Previously, we have established that the EHDI-induced ULH alignment is determined by the transport of mobile ions and, thus, the strength of space charge polarization unraveled in the complex dielectric spectra. Using dielectric spectroscopy, an optimized frequency range between f_L and f_R to facilitate EHDI-induced ULH has also been identified [18]. It is worth to mention that f_L in Figure 6 is the lower frequency at which the ε' and ε'' curves intersect each other and the relaxation frequency f_R ($>f_L$) refers to the one corresponding to the maximum value of the loss-tangent. As shown in Figure 6a, the value of ε' in the measured frequency regime (20 Hz–100 kHz), ascribed to the molecular orientation, is nearly constant. Moreover, based on this role, it is reasonable for the absence of ULH alignment in the cell applied with 30 Hz voltages at $T = 25$ °C in Figure 1a because both f_L and f_R fall lower than 20 Hz, the lower frequency limit of the LCR meter E4980A. In contrast, as the temperature increases to 50 °C, the three curves shift to higher frequencies due to the promoted ionic effect in the cell. As a consequence, f_L and f_R become 20 Hz and 63.2 Hz at 50 °C, respectively, so that the EHDI-induced ULH alignment can be acquired by the pulse of $V = 30$ V_{rms} at $f_2 = 30$ Hz (note: $f_L < f_2 < f_R$). This implies that when a CLC with low ion density is used, the ULH can be alternatively realized via the elevation in temperature and thus the resulting ionic effect by adjusting the frequency conditions of the hybrid voltage pulse to onset the EHDI.

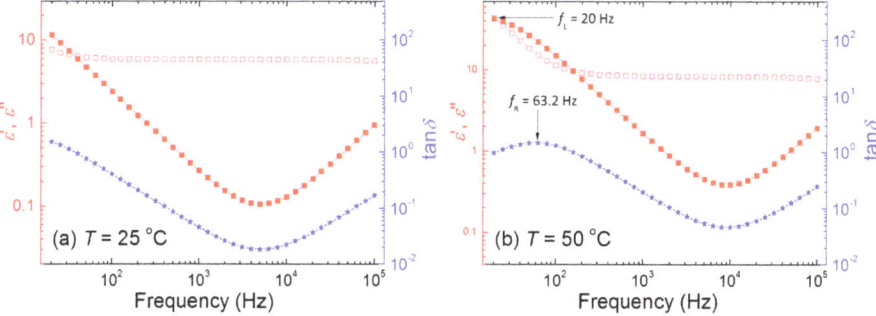

Figure 6. Complex dielectric and loss-tangent (tan δ) spectra of the CLC cell in the P state at temperatures of (a) $T = 25$ °C and (b) $T = 50$ °C.

4. Conclusions

Conventional methods to obtain a well-aligned ULH texture in a CLC cell with a simple surface-alignment condition (e.g., planar or hybrid) rely on the assisting voltage applied during the isotropic-to-CLC phase transition or via the EHDI. While the former requires a heating instrument in conjunction with a temperature controller to enable the phase transition so that the ULH can hardly switch to other CLC textures, the latter has been demonstrated specific for ion-rich positive CLCs. In the present work, we have referred to these two approaches and proposed to create ULH by applying a hybrid voltage across a CLC cell. The CLC used was injected into a 90°-twisted planar-aligned cell with nonzero electrode resistivity to enable dielectric heating via the pseudo-dielectric relaxation from the cell geometry [29].

With a hybrid voltage, the dielectric heating mechanism, which is dominated by the pseudo-dielectric relaxation originating from the ITO effect, permits the textural switching and phase transition in a CLC cell. On this basis under optimized voltage conditions, we have verified, by means of textural observations and measurements of time-dependent transmission and temperature, that the ULH alignment can be electrically generated through the thermal induction of the isotropic-to-CLC phase transition or by electrohydrodynamic instability in the homeotropic-to-CLC textural transition at a higher temperature to ensure a more pronounced ionic effect. The ULH obtained is switchable to another CLC texture, say, the Grandjean planar or focal conic state, by using the general scheme for the CLC textural switching. In contrast to the existing methods for the ULH formation reported in the literature, the proposed one presented in this study can adopt any CLC cell with simple surface alignment conditions (e.g., planar, 90°-twisted, and hybrid) and it is more universally applicable because ion-rich LCs and additional equipment for temperature controlling or mechanical shearing are not necessitated. Such an alternative approach holds great promise for overcoming the general hurdle of good-quality ULH, thereby facilitating its practical application for electro-optical devices.

Author Contributions: C.-H.Y. performed the experiments and drafted the manuscript. P.-C.W. conceived the experiments and helped analyze the data. Wei Lee supervised the whole study and finalized the manuscript.

Funding: This work was financially supported by the Ministry of Science and Technology, Taiwan, under grant Nos. 106-2923-M-009-002-MY3 and 107-2112-M-009-012-MY3.

Conflicts of Interest: The authors declare no conflict of interest.

References

1. Chilaya, G. Cholesteric Liquid Crystals: Optics, Electro-optics, and Photo-optics. In *Chirality in Liquid Crystals*; Kitzerow, H., Bahr, C., Eds.; Springer: New York, NY, USA, 2001; Chapter 6; pp. 159–185.
2. Huang, J.-C.; Hsiao, Y.-C.; Lin, Y.-T.; Lee, C.-R.; Lee, W. Electrically switchable organo–inorganic hybrid for a white-light laser source. *Sci. Rep.* **2016**, *6*, 28363. [CrossRef] [PubMed]
3. Hsiao, Y.-C.; Tang, C.-Y.; Lee, W. Fast-switching bistable cholesteric intensity modulator. *Opt. Express* **2011**, *19*, 9744–9749. [CrossRef]
4. White, T.J.; McConney, M.E.; Bunning, T.J. Dynamic color in stimuli-responsive cholesteric liquid crystals. *J. Mater. Chem.* **2016**, *20*, 9832–9847. [CrossRef]
5. Rudquist, P.; Komitov, L.; Lagerwall, S.T. Linear electro-optic effect in a cholesteric liquid crystal. *Phys. Rev. E* **1994**, *50*, 4735–4743. [CrossRef]
6. Musgrave, B.; Lehmann, P.; Coles, H.J. A new series of chiral nematic bimesogens for the flexoelectro-optic effect. *Liq. Cryst.* **1999**, *26*, 1235–1249. [CrossRef]
7. Coles, H.J.; Clarke, M.J.; Morris, S.M.; Broughton, B.J.; Blatch, A.E. Strong flexoelectric behavior in bimesogenic liquid crystals. *J. Appl. Phys.* **2006**, *99*, 034104. [CrossRef]
8. Morris, S.M.; Clarke, M.J.; Blatch, A.E.; Coles, H.J. Structure-flexoelastic properties of bimesogenic liquid crystals. *Phys. Rev. E* **2007**, *75*, 041701. [CrossRef]
9. Outram, B.I.; Elston, S.J. Frequency-dependent dielectric contribution of flexoelectricity allowing control of state switching in helicoidal liquid crystals. *Phys. Rev. E* **2013**, *88*, 012506. [CrossRef] [PubMed]
10. Varanytsia, A.; Chien, L.-C. Bimesogen-enhanced flexoelectro-optic behavior of polymer stabilized cholesteric liquid crystal. *J. Appl. Phys.* **2016**, *11*, 014502. [CrossRef]
11. Tan, G.; Lee, Y.-H.; Gou, F.; Hu, M.; Lan, Y.-F.; Tsai, C.-Y.; Wu, S.-T. Macroscopic model for analyzing the electro-optics of uniform lying helix cholesteric liquid crystals. *J. Appl. Phys* **2017**, *121*, 173102. [CrossRef]

12. Patel, J.S.; Meyer, R.B. Flexoelectric electro-optics of a cholesteric liquid crystal. *Phys. Rev. Lett.* **1987**, *58*, 1538–1540. [CrossRef] [PubMed]
13. Salter, P.S.; Elston, S.J.; Raynes, P.; Parry-Jones, L.A. Alignment of the uniform lying helix structure in cholesteric liquid crystals. *Jpn. J. Appl. Phys.* **2009**, *48*, 101302. [CrossRef]
14. Rudquist, P.; Komitov, L.; Lagerwall, S.T. Volume-stabilized ULH structure for the flexoelectro-optic effect and the phase-shift effect in cholesterics. *Liq. Cryst.* **1998**, *24*, 329–334. [CrossRef]
15. Inoue, Y.; Moritake, H. Discovery of a transiently separable high-speed response component in cholesteric liquid crystals with a uniform lying helix. *Appl. Phys. Express* **2015**, *8*, 061701. [CrossRef]
16. Inoue, Y.; Moritake, H. Formation of a defect-free uniform lying helix in a thick cholesteric liquid crystal cell. *Appl. Phys. Express* **2015**, *8*, 071701. [CrossRef]
17. Wang, C.-T.; Wang, W.-Y.; Lin, T.-H. A stable and switchable uniform lying helix structure in cholesteric liquid crystals. *Appl. Phys. Lett.* **2011**, *99*, 041108. [CrossRef]
18. Nian, Y.-L.; Wu, P.-C.; Lee, W. Optimized frequency regime for the electrohydrodynamic induction of a uniformly lying helix structure. *Photonics Res.* **2016**, *4*, 227–232. [CrossRef]
19. Yu, C.-H.; Wu, P.-C.; Lee, W. Alternative generation of well-aligned uniform lying helix texture in a cholesteric liquid crystal cell. *AIP Adv.* **2017**, *7*, 105107. [CrossRef]
20. Park, K.-S.; Baek, J.-H.; Lee, Y.-J.; Kim, J.-H.; Yu, C.-J. Effects of pretilt angle and anchoring energy on alignment of uniformly lying helix mode. *Liq. Cry.* **2016**, *43*, 1184–1189. [CrossRef]
21. Gardiner, D.J.; Morris, S.M.; Hands, P.J.W.; Castles, F.; Qasim, M.M.; Kim, W.-S.; Choi, S.S.; Wikinson, T.D.; Coles, H.J. Spontaneous induction of the uniform lying helix alignment in bimesogenic liquid crystals for the flexoelectro-optic effect. *Appl. Phys. Lett.* **2012**, *100*, 063501. [CrossRef]
22. Li, C.-C.; Tseng, H.-Y.; Chen, C.-W.; Wang, C.-T.; Jau, H.-C.; Wu, Y.-C.; Hsu, W.-H.; Lin, T.-H. Tri-stable cholesteric liquid crystal smart window. *SID DIGEST* **2018**, *49*, 543–545. [CrossRef]
23. Hegde, G.; Komitov, L. Periodic anchoring condition for alignment of a short pitch cholesteric liquid crystal in uniform lying helix texture. *Appl. Phys. Lett.* **2010**, *96*, 113503. [CrossRef]
24. Outram, B.I.; Elston, S.J. Spontaneous and stable uniform lying helix liquid-crystal alignment. *J. Appl. Phys.* **2013**, *113*, 043103. [CrossRef]
25. Komitov, L.; Brown, G.P.B.; Wood, E.L.; Smout, A.B.J. Alignment of cholesteric liquid crystals using periodic anchoring. *J. Appl. Phys.* **1999**, *86*, 3508–3511. [CrossRef]
26. Carbone, G.; Salter, P.; Elston, S.J.; Raynes, P.; De Sio, L.; Ferjani, S.; Strangi, G.; Umeton, C.; Bartolino, R. Short pitch cholesteric electro-optical device based on periodic polymer structures. *Appl. Phys. Lett.* **2009**, *9*, 011102. [CrossRef]
27. Outram, B.I.; Elston, S.J.; Tuffin, R.; Siemianowski, S.; Snow, B. The use of mould-templated surface structures for high-quality uniform-lying-helix liquid-crystal alignment. *J. Appl. Phys.* **2013**, *113*, 213111. [CrossRef]
28. Carbone, G.; Corbett, D.; Elston, S.J.; Raynes, P.; Jesacher, A.; Simmonds, R.; Booth, M. Uniform lying helix alignment on periodic surface relief structure generated via laser scanning lithography. *Mol. Cryst. Liq. Cryst.* **2011**, *544*, 37–49. [CrossRef]
29. Wu, P.-C.; Wu, G.-W.; Timofeev, I.V.; Zyryanov, V. Ya.; Lee, W. Electro-thermally tunable reflective colors in a self-organized cholesteric helical superstructure. *Photonics Res.* **2018**, *6*, 1094–1100. [CrossRef]
30. Wu, P.-C.; Hsiao, C.-Y.; Lee, W. Photonic bandgap–cholesteric device with electrical tunability and optical tristability in its defect modes. *Crystals* **2017**, *7*, 184. [CrossRef]
31. Schadt, M. Dielectric heating and relaxations in nematic liquid crystals. *Mol. Cryst. Liq. Cryst.* **1981**, *66*, 319–336. [CrossRef]
32. Jian, B.-R.; Tang, C.-Y.; Lee, W. Temperature-dependent electrical properties of dilute suspensions of carbon nanotubes in nematic liquid crystals. *Carbon* **2011**, *49*, 910–914. [CrossRef]

© 2019 by the authors. Licensee MDPI, Basel, Switzerland. This article is an open access article distributed under the terms and conditions of the Creative Commons Attribution (CC BY) license (http://creativecommons.org/licenses/by/4.0/).

Article

Effect of Size Polydispersity on the Pitch of Nanorod Cholesterics

Henricus H. Wensink

Laboratoire de Physique des Solides—UMR 8502, CNRS, Université Paris-Sud, Université Paris-Saclay, 91405 Orsay, France; rik.wensink@u-psud.fr

Received: 13 February 2019; Accepted: 4 March 2019; Published: 10 March 2019

Abstract: Many nanoparticle-based chiral liquid crystals are composed of polydisperse rod-shaped particles with considerable spread in size or shape, affecting the mesoscale chiral properties in, as yet, unknown ways. Using an algebraic interpretation of Onsager-Straley theory for twisted nematics, we investigate the role of length polydispersity on the pitch of nanorod-based cholesterics with a continuous length polydispersity, and find that polydispersity enhances the twist elastic modulus, K_2, of the cholesteric material without affecting the effective helical amplitude, K_t. In addition, for the infinitely large average aspect ratios considered here, the dependence of the pitch on the overall rod concentration is completely unaffected by polydispersity. For a given concentration, the increase in twist elastic modulus (and reduction of the helical twist) may be up to 50% for strong size polydispersity, irrespective of the shape of the unimodal length distribution. We also demonstrate that the twist reduction is reinforced in bimodal distributions, obtained by doping a polydisperse cholesteric with very long rods. Finally, we identify a subtle, non-monotonic change of the pitch across the isotropic-cholesteric biphasic region.

Keywords: chirality; cholesterics; nanorods; Onsager theory; polydispersity

1. Introduction

Polydispersity is widespread in colloidal and polymeric systems, since the building blocks are never fully identical but exhibit a continuous spread in size, shape, or surface charge [1]. The variety in microscopic interactions ensuing from polydispersity may have a considerable influence on the phase stability [2,3] or the mechanical properties of nanoparticle-based materials through aggregation [4], packing [5], or percolation processes [6]. Research efforts can be aimed at either purifying colloidal suspensions in order to promote crystallization [7], such as through templating [8], or at deliberately enhancing size polydispersity; for instance, to improve the electronic conductivity of percolated rod networks [6,9], to stabilize glassy states of spherical particles [10,11], or to realize complex fluids with bespoke rheological properties [12]. The effect of size polydispersity in lyotropic liquid crystals composed of non-spherical (e.g., rod-shaped) nanoparticles was first addressed in the 1980s, focussing mostly on its effect on the nematic osmotic pressure [13], on the stability of smectic order [14], and on the impact of size bidispersity on the order-disorder transition [15,16].

The presence of chiral forces among rod-shaped particles is usually expressed in terms of some helical organization on the mesoscale, as is the case, for instance, in chiral nematics or cholesterics [17,18]. The helical twist of the local nematic director defines a typical mesoscopic lengthscale, referred to as the pitch, whose controllability is of key importance in the manifold examples of chiral nematics involved in technological applications (e.g., displays), as well as in nature [19]. Cholesteric materials based on nanorods commonly consist of rigid, fibrillar units, composed of some biological component such as cellulose (CNCs) [20–22], chitin [23,24], collagen [25], or amyloid [26,27]. These fibrils are inherently size-polydisperse and the effect of size disparity on the sensitivity of the pitch remains an important outstanding issue. In these systems, size polydispersity is quenched by the synthesis procedure and usually does not depend on the thermodynamic state of the system. Similar to chiral chromonics [28], nanometric chiral building blocks, such as short-fragment DNA [29,30], may reversibly polymerize into chiral filaments that are inherently polydisperse. However, these systems constitute a different class of cholesterics, characterized by annealed polydispersity where the contour length distribution of the filaments is dictated by temperature, the degree of semiflexibility, and the monomer concentration [31,32].

In this paper, we attempt to address the effect of quenched length polydispersity on cholesterics from a theoretical viewpoint, and propose an algebraic theory that is capable of linking the cholesteric pitch to the microscopic chirality of the rods, as well as their inherent length distribution. We find that length-polydispersity has a significant impact on the twist elastic modulus of the cholesteric material, increasing it by about 50% compared to its monodisperse counterpart at the same overall rod concentration. Within the same framework, we also address the isotropic-cholesteric phase coexistence and identify the concentration, length-composition, and pitch of the cholesteric phase fraction upon traversing the biphasic region, revealing subtle non-monotonic trends that could be exploited to purify or control the size composition of a cholesteric material. We hope that the present theory may serve as a useful tool in guiding or rationalizing certain experimental trends regarding the pitch of biofibrillar-based cholesteric systems with quenched length polydispersity.

2. Onsager-Straley Theory for Polydisperse Cholesterics

Let us start with the free energy per unit volume V of a polydisperse assembly of strongly elongated rods with diameter D and length L, the latter following some quenched length distribution $c(\ell)$ with renormalized rod length ℓ. Within Onsager's second-virial approximation [33], the free energy of the rod fluid per unit volume reads:

$$f = \frac{v_0 F}{V} \sim \int d\ell c(\ell)(\ln c(\ell) - 1) + \int d\ell c(\ell)\sigma^{(\ell)} + \iint d\ell d\ell' c(\ell)c(\ell')\left[\rho^{(\ell\ell')} + f_c^{(\ell\ell')}(q)\right], \quad (1)$$

where $\beta = (k_B T)^{-1}$ denotes the thermal energy in terms of Boltzmann's constant k_B and temperature T. The renormalized rod length $\ell = L/L_0$ exhibits a continuous spread prescribed by a normalised distribution $p(\ell)$, so that $c(\ell) = c_0 p(\ell)$ in terms of the overall dimensionless particle density $c_0 = Nv_0/V$ and microscopic volume $v_0 = \pi L_0^2 D/4$, with L_0 the average rod length. Consequently, the first moment of the distribution is fixed at unity (i.e., $\int d\ell p(\ell)\ell = 1$).

The free energy consists of three entropic contributions relating to the ideal gas, orientational, and excluded volume entropy, respectively. The first two entropic quantities can be computed in their exact form, while the excluded-volume entropy is defined on the level of the second-virial coefficient between a pair of rods. This approximation should be accurate if all rod species are sufficiently slender and that their aspect ratio $L/D \gg 1$ [33]. The orientational entropy is defined as:

$$\sigma^{(\ell)} = \int d\Omega \psi(\Omega, \ell)[4\pi\psi(\Omega, \ell)], \quad (2)$$

and involves some unknown orientational distribution function $\psi(\Omega, \ell)$ that describes the orientational probability of a rod with length ℓ in terms of a solid angle Ω. Trivially, for an isotropic fluid, where the rods point in random directions, the distribution becomes a mere constant $\psi = (4\pi)^{-1}$, irrespective of ℓ. The orientational entropic factor is then simply rendered zero (i.e., $\sigma = 0$).

The second entropic contribution ρ is defined as the angular-averaged excluded volume per particle in a nematic phase, normalized to its random isotropic average:

$$\rho^{(\ell,\ell')} = \frac{4}{\pi}\ell\ell' \iint d\Omega d\Omega' \psi(\Omega, \ell)\psi(\Omega', \ell')|\sin \gamma|, \quad (3)$$

with γ denoting the enclosed angle between two rods (see Figure 1). For the isotropic phase, it is easily established that $\langle\langle|\sin\gamma|\rangle\rangle_\psi = \pi/4$ and $\rho^{(\ell\ell')} = \ell\ell'$.

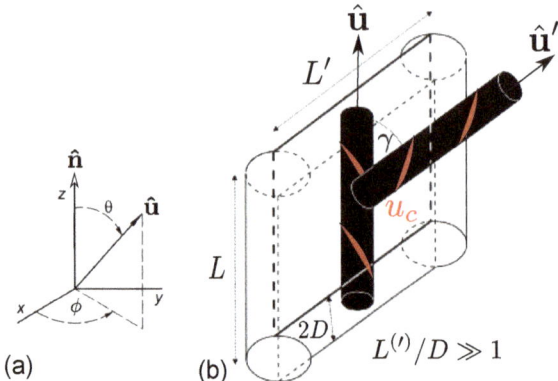

Figure 1. (a) Overview of the lab frame, nematic director $\hat{n}$, and principal angles used in the present analysis; (b) Excluded volume between two achiral hard cylinders decorated with a perturbative chiral potential u_c acting locally along the rod contour (indicated by the red helical threads). The rod excluded volume is assumed to be unaffected by the chiral potential and is responsible for stabilizing the nematic order (ρ) and generating twist elasticity (K_2), while the chiral potential u_c promotes director twist (K_t).

The last contribution in Equation (1) is due to Straley [34], and describes the free energy difference between the weakly twisted director field of a cholesteric liquid crystal and the uniform one of a nematic. The degree of helical organization is defined in terms of a dimensionless wave number, $q = 2\pi L_0/p_c$, where the pitch is required to be much larger than the average nanoparticle size (i.e., $p_c \gg L_0$). Under these restrictions, $q \ll 1$, and the additional free energy density takes a simple quadratic form:

$$f_c^{(\ell\ell')}(q) = qK_t^{(\ell\ell')} + \frac{1}{2}q^2 K_2^{(\ell\ell')}, \tag{4}$$

in terms of a species-dependent helical amplitude $K_t^{(\ell\ell')}$ and twist elastic modulus $K_2^{(\ell\ell')}$, defined microscopically as:

$$\begin{aligned} K_t^{(\ell\ell')} &\sim \iint d\Omega d\Omega' \psi(\Omega,\ell)\dot\psi(\Omega',\ell')\Omega'_\perp M_t^{(\ell\ell')}(\Omega,\Omega'), \\ K_2^{(\ell\ell')} &\sim \iint d\Omega d\Omega' \dot\psi(\Omega,\ell)\dot\psi(\Omega',\ell')\Omega_\perp \Omega'_\perp M_2^{(\ell\ell')}(\Omega,\Omega'). \end{aligned} \tag{5}$$

These expressions depend on the derivative of the local orientational probability $\dot\psi = \partial\psi/\partial\Omega$ with $\Omega_\perp$ denoting the component of the rod orientation perpendicular to the local nematic director and the pitch axis. The kernels describe the interactions between the chiral rods, which we assume to consist of a weak soft potential u_c imparting chirality superimposed onto a hard-core repulsion generated by the cylindrical backbone that is responsible for generating twist elasticity. The helical amplitude M_t is given by an integrated (van der Waals) potential [35,36]:

$$M_t^{(\ell\ell')}(\Omega,\Omega') \sim (v_0 L_0)^{-1} \int_{\notin v_{\text{excl}}} d\mathbf{r}_\parallel \beta u_c^{(\ell\ell')}(\mathbf{r},\Omega,\Omega'), \tag{6}$$

and depends uniquely on the chiral potential $u_c^{(\ell\ell')}$ between rods of length ℓ and ℓ', which we will specify shortly. Here, $\mathbf{r}_\parallel$ represents the component of the centre-of-mass distance between a rod pair along the pitch axis. The second kernel, M_2, depends on a generalized excluded-volume between the achiral cylindrical backbone of two rods of different lengths, and reads for slender rods [36,37]:

$$M_2^{(\ell\ell')}(\Omega,\Omega') \sim -\frac{2}{3\pi}\ell\ell'|\sin\gamma|(\ell^2\Omega_\parallel^2 + \ell'^2\Omega_\parallel'^2), \tag{7}$$

where $\Omega_\parallel$ is the rod orientation projected along the pitch axis.

Although the twisting of the director changes the local uniaxial alignment in favour of biaxial order [38], the biaxial perturbation is very weak for $q \ll 1$, and we shall assume that the local uniaxial nematic order remains unperturbed. Consequently, the orientational distribution depends solely on the polar angle, θ, between the main particle orientation vector, $\hat{u}$, and the nematic director, $\hat{n}$, by $\cos\theta = \hat{u}\cdot\hat{n}$. Let us further assume strongly nematic order, so that the use of a Gaussian Ansatz [13,39] for the local orientational probability is justified:

$$\psi_G(\theta,\ell) \sim \frac{\alpha(\ell)}{4\pi}\exp\left(-\frac{1}{2}\alpha(\ell)\theta^2\right), \tag{8}$$

supplemented with its polar mirror form $\psi(\pi-\theta,\ell)$ along $-\hat{n}$, in order to guarantee local apolar order. The variational parameter $\alpha(\ell)$ is required to be much larger than unity and is length-dependent. While $\alpha(\ell)$ is, as yet, unknown in explicit form, common sense tells us that $\alpha(\ell)$ should be proportional to the rod contour length, since long rods tend to be more strongly aligned than short rods [13,15]. The Gaussian approximation cannot represent isotropic order since, upon taking $\alpha\downarrow 0$, the expression above reduces to zero, rather than giving the desired form $\psi = 1/4\pi$. There are consistent algebraic trial functions for ψ that do correctly render isotropic order in this limit, but these involve more complicated distributions that tend to compromise the tractability of the theory [33,40].

From Equation (8) we readily find an asymptotic expression for the orientational entropic factor:

$$\sigma^{(\ell)} \sim \ln\alpha(\ell) - 1, \tag{9}$$

whereas the excluded-volume term can be estimated from an asymptotic expansion for $\alpha \gg 1$ giving up to the leading order [39]:

$$\rho^{(\ell,\ell')} \sim \frac{4}{\pi}\ell\ell'\left(\frac{\pi}{2}\right)^{\frac{1}{2}}\left(\frac{1}{\alpha(\ell)} + \frac{1}{\alpha(\ell')}\right)^{\frac{1}{2}}. \tag{10}$$

Since the effective torque associated with director twist is relatively weak compared to the one enforcing nematic order, it is safe to assume that $\alpha(\ell)$ does not depend on the pitch. Using the results of Equations (2) and (3) in the free energy of the untwisted nematic, Equation (1) (with $q=0$) enables a formal minimization with respect to $\alpha(\ell)$, giving the following self-consistency condition:

$$\tilde{\alpha}^{\frac{1}{2}}(\ell) = 2^{\frac{1}{2}}\int d\ell'\ell\ell' p(\ell')g_0(\ell,\ell'), \tag{11}$$

with

$$g_0(\ell,\ell') = \left(1+\frac{\tilde{\alpha}(\ell)}{\tilde{\alpha}(\ell')}\right)^{-\frac{1}{2}}. \tag{12}$$

No matter what length distribution, $\alpha(\ell)$ scales quadratically with concentration c_0, so that it is expedient to factorize

$$\alpha(\ell) = \frac{4}{\pi}c_0^2\tilde{\alpha}(\ell), \tag{13}$$

with $\tilde{\alpha}(\ell)$ depending only on the shape of the normalised distribution $p(\ell)$. Unfortunately, Equation (11) does not permit $\alpha(\ell)$ to be resolved analytically, but a numerical solution is easily obtained for a given distribution $p(\ell)$ [41].

Minimizing Equation (1) with respect to q, we obtain the equilibrium value for the wave number q reflecting a balance between the helical amplitude and twist elastic modulus:

$$q \equiv \frac{K_t}{K_2} = -\frac{\iint d\ell d\ell' c(\ell) c(\ell') K_t^{(\ell\ell')}}{\iint d\ell d\ell' c(\ell) c(\ell') K_2^{(\ell\ell')}}. \tag{14}$$

These contributions will be computed in algebraic form in the next Section.

3. Asymptotic Results for the Helical Amplitude and Twist Elastic Modulus

Let us now propose a simple chiral potential acting between two freely rotating rods. We shall consider the commonly used pseudo-scalar form [42,43]:

$$u_c^{(\ell\ell')}(\mathbf{r}, \Omega, \Omega') \sim \varepsilon g(r) \left(\hat{\mathbf{u}}_1 \times \hat{\mathbf{u}}_2 \cdot \hat{\mathbf{r}} \right), \tag{15}$$

with $g(r)$ some rapidly decaying function of the centre-of-mass distance r and ε specifying the microscopic chiral strength between the rods. We may work out the integrated chiral potential M_t corresponding to this potential, first by defining the pitch axis of the cholesteric to align along the x-axis of a Cartesian laboratory frame (see Figure 1), and defining a rod orientation $\hat{\mathbf{u}} = (\sin\theta \sin\varphi, \sin\theta \cos\varphi, \cos\theta)$ in terms of polar and azimuthal angles (θ, φ), with respect to a reference nematic director $\hat{\mathbf{n}}$ pointing along the z-axis. Then, $\Omega_\perp = u_y$ and we may perform a Taylor expansion for $\theta \ll 1$ and keep only the leading order contribution. Some algebraic manipulations, along the lines proposed in [37,44], lead to the following asymptotic expression:

$$\Omega'_\perp M_t^{(\ell\ell')} \sim \bar{\varepsilon} \ell \ell' \left[(\theta'^2 - \theta^2) + |\gamma|^2 \right], \tag{16}$$

with $\bar{\varepsilon}$ a dimensionless chiral strength combining various microscopic features:

$$\bar{\varepsilon} \sim \frac{1}{\pi} \beta \varepsilon \frac{D}{L_0} \int_D^\infty dr r g(r). \tag{17}$$

A similar analysis can be performed for the twist elastic contribution M_2 producing the following angular dependency for strong alignment [44]:

$$\Omega_\perp \Omega'_\perp M_2^{(\ell\ell')} \sim -\frac{\ell\ell'}{24\pi} \left[|\gamma|(\theta'^2 + \theta^2)(\ell'^2 \theta'^2 + \ell^2 \theta^2) - |\gamma|^3 (\ell'^2 \theta'^2 + \ell^2 \theta^2) \right]. \tag{18}$$

The remaining task is to perform Gaussian orientational averages of these quantities, as per Equation (5), to arrive at an explicit expression for the kernels M_t and M_2. The mathematical theorem that allows one to compute the angular averages has been discussed in Onsager's original paper [33], and used later on in Odijk's work on elastic constants [44]. The averages are given in explicit form in Appendix A. An additional advantage of the Gaussian approach is that we can use the simple relation $\dot{\psi}_G \sim \alpha(\ell) \psi_G$ to obtain the derivate of the orientational distributions involved in Equation (5). Straightforward algebraic manipulation then leads to a simple result for the helical amplitude:

$$K_t^{(\ell\ell')} \sim 2\bar{\varepsilon} \ell \ell'. \tag{19}$$

The overall helical amplitude is *independent* of the length distribution, and scales quadratically with rod concentration c_0:

$$K_t \sim 2c_0^2 \bar{\varepsilon}. \tag{20}$$

We remark that this result may be different for purely steric chirality induced by some helical nanorod shape, such as a corkscrew [45–47]. More complicated chiral interactions—for example, those generated by a helical arrangement of charged surface groups, as in the case of viral rods [48]—can, in principle, be captured within a numerical interpretation of the van der Waals term Equation (6). The twist elastic modulus for a polydisperse nematic takes on a more elaborate form:

$$K_2 \sim \frac{c_0}{12\pi 2^{\frac{1}{2}}} \iint d\ell d\ell' p(\ell) p(\ell') \ell\ell' \left(\frac{1}{\tilde{a}(\ell)} + \frac{1}{\tilde{a}(\ell')} \right)^{\frac{1}{2}} \frac{\ell^2 \tilde{a}(\ell')[4\tilde{a}(\ell)+3\tilde{a}(\ell')]+\ell'^2 \tilde{a}(\ell)[3\tilde{a}(\ell)+4\tilde{a}(\ell')]}{[\tilde{a}(\ell)+\tilde{a}(\ell')]^2}. \quad (21)$$

For monodisperse sytems, when $\alpha(\ell) = \alpha(\ell') = \alpha$ and $\ell = \ell' = 1$, one recovers Odijk's scaling result $K_2 = c_0^2 K_2^{(11)} \sim 7c_0/24\pi$ [44]. From Equation (14), we infer that the cholesteric pitch always decreases with overall rod concentration through $p_c \sim q^{-1} \propto c_0^{-1}$, irrespective of polydispersity. The length distribution will, of course, have an effect on the pitch, but only through modification of the twist elasticity, as evident from Equation (21). We will explore this in more detail in the next Section.

4. Results for Log-Normal and Schulz-Distributed Rod Lengths

A typical size distribution for polymers [49], as well as for colloidal particles with quenched polydispersity [50], is the log-normal distribution, which is based on the logarithm of the rod length following a normal distribution:

$$p(\ell) = \frac{1}{(2\pi)^{\frac{1}{2}} w \ell} \exp\left[-\frac{(\ln \ell + \frac{w^2}{2})^2}{2w^2}\right], \quad (22)$$

with natural bounds $\ell_{min} = 0$ and $\ell_{max} \to \infty$. Equation (22) has unity mean $\langle \ell \rangle = 1$, whereas the polydispersity σ is connected to the standard deviation by:

$$\sigma = \left(\frac{\langle \ell^2 \rangle - \langle \ell \rangle^2}{\langle \ell \rangle^2} \right)^{\frac{1}{2}}, \quad (23)$$

through $\sigma^2 = e^{w^2} - 1$. Finite-tail cutoffs lead to small corrections that are easily accounted for numerically. Alternatively, a commonly-used form representing polymer molecular weight distributions is the Schulz-Zimm function [51,52]:

$$p(\ell) = \frac{(1+z)^{1+z}}{\Gamma(1+z)} \ell^z \exp(-(z+1)\ell), \quad (24)$$

which is normalized on the domain $0 < \ell < \infty$ and has mean $\langle \ell \rangle = 1$ and polydispersity $\sigma = (1+z)^{-1/2}$. The exponential tail renders cut-off effects far less serious than for the log-normal form [53]. The results for both distributions are shown in Figure 2. For the log-normal distribution cut-off values of $\ell_{min} = 0.01$ and $\ell_{max} = 20$ were used. The increase of the twist elastic modulus with polydispersity σ appears significant and robust, as it is mostly insensitive to the shape of the length distribution and the cut-off values. Clearly, introducing a spread of rod lengths at a given overall concentration induced a significant "stiffening" of the nematic fluid with respect to a twist distortion of the director.

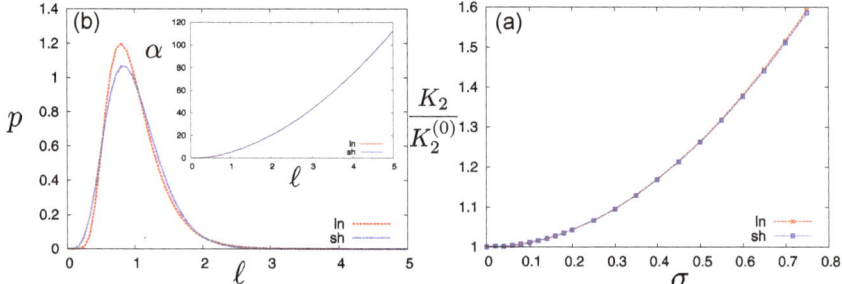

Figure 2. (a) Overview of log-normal and Schulz-length distributions with polydispersity $\sigma = 0.4$. The inset depicts the associated Gaussian parameter a versus rod length ℓ, obtained from Equation (11), showing that long rods align much more strongly than short ones; (b) Twist elastic modulus K_2 for a polydisperse nanorod cholesteric with increasing polydispersity σ, normalized to the value $K_2^{(0)}$ of the corresponding monodisperse system. The results for both distributions are virtually indistinguishable.

4.1. Effect of Large-Rod Dopants and Bimodality

We will now explore the effect of doping a unimodally length-distributed nanorod cholesteric with a tiny fraction of large rods of length ℓ_{max}. Let us supplement the log-normal distribution Equation (22) with a growing exponential tail to construct a weakly bimodal size distribution [54]:

$$p_d(\ell) \sim p(\ell) + xe^{-a(\ell_{max} - \ell)}, \qquad (25)$$

where $x \ll 1$ is the mole fraction of the added rods and $a \gg 1$ quantifies the degree of bimodality. Equation (25) lacks a trivial normalization factor, which is included in the numerical calculations. Moreover, the alteration of the log-normal parent distribution affects the renormalized average length, such that $\langle \ell \rangle = \int d\ell \ell p(\ell) > 1$ which, in turn, changes the helical amplitude K_t through Equation (19). These effects are easily accounted for numerically. Results for $a = 10$ and a unimodal polydispersity σ of 30% (which seems a typical value, for example, for CNCs [22]) are shown in Figure 3, for different values of the maximum cut-off ℓ_{max}. The results demonstrate that adding even a very small fraction of long rods (less than 1%) causes a significant reduction of the helical twist. The strength of the reduction can be systematically tuned through the length of the doped rods. Since the bimodal twist reduction of the cholesteric system is imparted mostly through modification of its twist elasticity, the microscopic chiral properties of the doped rods are not imminently important provided their number fraction remains sufficiently small.

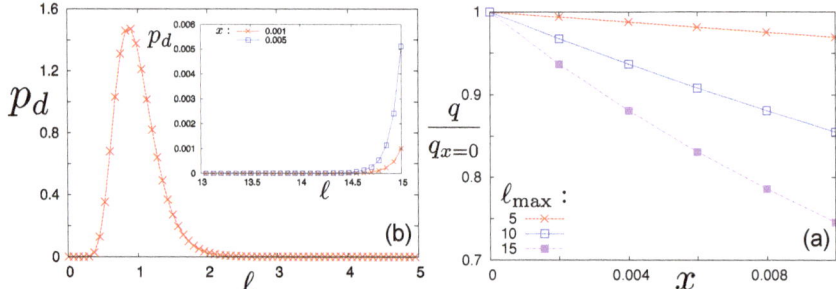

Figure 3. (a) Overview of a weakly bimodal log-normal distribution ($\sigma = 0.3$) with an exponential tail at $\ell_{max} = 15$ at different mole fractions x of long-rod dopants (inset); (b) Reduction of the cholesteric pitch wave number q, upon increasing the mole fraction x of large rods with ℓ_{max} times the average rod length.

4.2. Pitch Variation across the Isotropic-Cholesteric Biphasic Region

We finish our analysis by investigating the behaviour of the cholesteric pitch within the isotropic-nematic biphasic region. The thermodynamics of phase transitions of length-polydisperse rod systems within the Gaussian Ansatz has been discussed in detail, in [41]. We may determine coexistence between the isotropic and cholesteric phases by imposing equality of osmotic pressure and chemical potential, both of which are straightforward derivatives of the free energy Equation (1). At finite phase fractions, the distribution of rod lengths in each of the coexisting phases is different from the imposed log-normal parent distribution. Concomitantly, the cholesteric pitch will depend non-trivially on the phase fraction, or the location within the biphasic region. This is illustrated in Figure 4, showing the variation of the pitch as well as the evolution of the concentration and polydispersity of the cholesteric phase fraction across the biphasic region. Upon moving away from the isotropic-cholesteric (I–C) cloud point ($x_{\text{chol}} = 0$), where only a infintesimal fraction of cholesteric phase has been formed (referred to as the "shadow"), the cholesteric unwinds initially and then rewinds (i.e., tighter pitch lengths) close to the C–I cloud point ($x_{\text{chol}} = 1$). In the latter, where a negligible fraction of isotropic phase is left, the length distribution within the cholesteric phase equals the log-normal parental one. The non-monotonic trend of the pitch is not inflicted by the cholesteric concentration, which increases gradually upon x_{chol}, but is the result of subtle changes in the length variation upon traversing the biphasic region. We remark that the polydispersity of the cholesteric phase is at its lowest (about $\sigma \approx 0.3$) at the I–C cloud point, thus offering a simple means for purifying a polydisperse cholesteric system through successive sweeps of phase separation. In addition, splitting off the cholesteric phase fraction close to the I–C cloud point provides an effective way of "filtering out" the largest rods, given that the average rod length is larger than overall, as suggested by the distribution for $x_{\text{chol}} = 0.01$ in the inset of Figure 4.

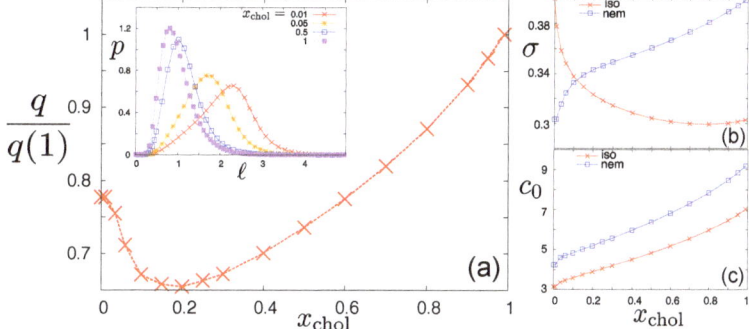

Figure 4. (a) Variation of the pitch across the isotropic-cholesteric (I–C) biphasic region for a nanorod system having a log-normal length distribution with $\sigma = 0.4$. Plotted is the pitch, q, renormalized to its value $q(1)$ at the C–I cloud point, versus the cholesteric phase fraction, x_{chol}. The inset depicts a number of length distributions in the cholesteric phase at different phase fractions; The panels (b) and (c), on the right, indicate the polydispersity, σ, and the concentration, c_0; (c) of the coexisting phases versus x_{chol}.

5. Conclusions

Inspired by a recent upsurge in experimental studies on cholesteric self-organization of rigid chiral nanorods with quenched length polydispersity (most notably, microfibrils made of cellulose [22], chitin [23,24], and related biocomponents) we have extended the Onsager-Straley theory [33,34] for the cholesteric organization of chiral rods with uniform length, towards the polydisperse case. The central assumptions underlying the theoretical analysis are the following: (i) The rods are completely rigid and sufficiently slender, so as to respect the Onsager limit of infinite length-to-width ratio; (ii) the local nematic alignment along the revolving director describing a twisted nematic is asymptotically strong, which justifies the use of a simple Gaussian variational approach [39] for the local orientational probability; and (iii) the helical deformation, q, of the director field is weak, on the scale of the average rod length L_0, so that $qL_0 \gg 1$. We show that, with these criteria fulfilled, the Onsager-Straley theory can be cast in an algebraic form. The determination of the pitch for a given length distribution requires relatively little computational cost, save for a straighforward numerical iterative procedure to determine the length-dependence of the variational parameter describing the degree of nematic order. Our main finding is that length polydispersity principally enhances the twist elasticity of a cholesteric material, with the helical twisting power (generated by the microscopic chirality of the rods) being only marginally affected. Quantitative examples are given of a pitch reduction generated by doping a polydisperse cholesteric system with long rods residing in the tail of the unimodal length distribution.

Without claiming to have presented an accurate theory for any chiral nanorod assembly in particular, we believe the present algebraic theory to be capable of providing a tractable and physically insightful tool that may be helpful for interpreting and guiding experimental observations in these systems. In particular, our findings demonstrate that the isotropic-cholesteric phase transition can be used as a useful vehicle to purify or select chiral species of a certain length, or to fine-tune the pitch of polydisperse nanorod cholesterics.

Conflicts of Interest: The author declare no conflict of interest.

Appendix A. Gaussian Averages

The procedure for obtaining Gaussian averages needed for the computation of the twist elasticity $K_2^{(\ell\ell')}$ of a polydisperse nematic is given in Reference [44]. The following averages are required:

$$\langle\langle |\gamma|^3 \theta^2 \rangle\rangle_{\psi_G} \sim 3 \left(\frac{\pi}{2}\right)^{\frac{1}{2}} \left(\frac{1}{\alpha_1} + \frac{1}{\alpha_2}\right)^{\frac{1}{2}} \frac{2\alpha_1 + 5\alpha_2}{\alpha_1^2 \alpha_2},$$

$$\langle\langle |\gamma| \theta^4 \rangle\rangle_{\psi_G} \sim \left(\frac{\pi}{2}\right)^{\frac{1}{2}} \left(\frac{1}{\alpha_1} + \frac{1}{\alpha_2}\right)^{\frac{1}{2}} \frac{8\alpha_1^2 + 24\alpha_1\alpha_2 + 15\alpha_2^2}{\alpha_1^2 (\alpha_1 + \alpha_2)^2}, \quad (A1)$$

$$\langle\langle |\gamma| \theta^2 \theta'^2 \rangle\rangle_{\psi_G} \sim \left(\frac{\pi}{2}\right)^{\frac{1}{2}} \left(\frac{1}{\alpha_1} + \frac{1}{\alpha_2}\right)^{\frac{1}{2}} \frac{6\alpha_1^2 + 11\alpha_1\alpha_2 + 6\alpha_2^2}{\alpha_1 \alpha_2 (\alpha_1 + \alpha_2)^2},$$

where we denote $\alpha_1 = \alpha(\ell)$ and $\alpha_2 = \alpha(\ell')$.

References

1. Fasolo, M.; Sollich, P.; Speranza, A. Phase equilibria in polydisperse colloidal systems. *React. Funct. Polym.* **2004**, *58*, 187–196. [CrossRef]
2. Huang, S.H.; Radosz, M. Equation of state for small, large, polydisperse and associating molecules—Extension to fluid mixtures. *Ind. Eng. Chem. Res.* **1991**, *30*, 1994. [CrossRef]
3. Sollich, P. Predicting phase equilibria in polydisperse systems. *J. Phys. Condens. Matter* **2002**, *14*, 79. [CrossRef]
4. Bushell, G.; Amal, R. Fractal aggregates of polydisperse particles. *J. Colloid Interface Sci.* **1998**, *205*, 459. [CrossRef] [PubMed]
5. Farr, R.S.; Groot, R.D. Close packing density of polydisperse hard spheres. *J. Chem. Phys.* **2009**, *131*, 244104. [CrossRef] [PubMed]
6. Kyrylyuk, A.V.; van der Schoot, P. Continuum percolation of carbon nanotubes in polymeric and colloidal media. *Proc. Natl. Acad. Sci. USA* **2008**, *105*, 8221–8226. [CrossRef] [PubMed]
7. Murray, C.B.; Kagan, C.R.; Bawendi, M.G. Synthesis and characterization of monodisperse nanocrystals and close-packed nanocrystal assemblies. *Annu. Rev. Mater. Sci.* **2000**, *30*, 545–610. [CrossRef]
8. Van Blaaderen, A.; Ruel, R.; Wiltzius, P. Template-directed colloidal crystallization. *Nature* **1997**, *385*, 321–324. [CrossRef]

9. Kyrylyuk, A.V.; Hermant, M.C.; Schilling, T.; Klumperman, B.; Koning, C.E.; van der Schoot, P. Controlling electrical percolation in multicomponent carbon nanotube dispersions. *Nat. Nanotechnol.* **2011**, *6*, 364. [CrossRef] [PubMed]
10. Auer, S.; Frenkel, D. Suppression of crystal nucleation in polydisperse colloids due to increase of the surface free energy. *Nature* **2001**, *413*, 711. [CrossRef] [PubMed]
11. Pusey, P.N.; Zaccarelli, E.; Valeriani, C.; Sanz, E.; Poon, W.C.K.; Cates, M.E. Hard spheres: Crystallization and glass formation. *Philos. Trans. A Math. Phys. Eng. Sci.* **2009**, *367*, 4993. [CrossRef] [PubMed]
12. Ten Brinke, A.J.W.; Bailey, L.; Lekkerkerker, H.N.W.; Maitland, G.C. Rheology modification in mixed shape colloidal dispersions. Part II: Mixtures. *Soft Matter* **2008**, *4*, 337–348. [CrossRef]
13. Odijk, T. Osmotic pressure of a nematic solution of polydisperse rod-like macromolecules. *Liq. Cryst.* **1986**, *1*, 97–100. [CrossRef]
14. Sluckin, T.J. Polydispersity in liquid crystal systems. *Liq. Cryst.* **1989**, *1*, 111–131. [CrossRef]
15. Odijk, T.; Lekkerkerker, H.N.W. Theory of the isotropic-liquid crystal phase separation for a solution of bidisperse rodllke macromolecules. *J. Phys. Chem.* **1985**, *89*, 2090–2096. [CrossRef]
16. Vroege, G.J.; Lekkerkerker, H.N.W. Phase transitions in lyotropic colloidal and polymer liquid crystals. *Rep. Prog. Phys.* **1992**, *55*, 1241. [CrossRef]
17. De Gennes, P.G.; Prost, J. *The Physics of Liquid Crystals*; Clarendon Press: Oxford, UK, 1993.
18. Dierking, I. Chiral Liquid Crystals: Structures, Phases, Effects. *Symmetry* **2014**, *6*, 444–472. [CrossRef]
19. Mitov, M. Cholesteric liquid crystals in living matter. *Soft Matter* **2017**, *13*, 4176–4209. [CrossRef] [PubMed]
20. Werbowyj, R.S.; Gray, D.G. Liquid Crystalline Structure In Aqueous Hydroxypropyl Cellulose Solutions. *Mol. Cryst. Liq. Cryst.* **1976**, *34*, 97–103. [CrossRef]
21. Heux, L.; Chauve, G.; Bonini, C. Nonflocculating and chiral-nematic self-ordering of cellulose microcrystals suspensions in nonpolar solvents. *Langmuir* **2000**, *16*, 8210–8212. [CrossRef]
22. Lagerwall, J.P.F.; Schütz, C.; Salajková, M.; Noh, J.; Park, J.H.; Scalia, G.; Bergström, L. Cellulose nanocrystal-based materials: From liquid crystal self-assembly and glass formation to multifunctional thin films. *NPG Asia Mater.* **2014**, *6*, e80. [CrossRef]
23. Revol, J.F.; Marchessault, R. In vitro chiral nematic ordering of chitin crystallites. *Int. J. Biol. Macromol.* **1993**, *15*, 329–335. [CrossRef]
24. Belamie, E.; Davidson, P.; Giraud-Guille, M.M. Structure and chirality of the nematic phase in α-chitin suspensions. *J. Phys. Chem. B* **2004**, *108*, 14991–15000. [CrossRef]
25. Giraud-Guille, M.M.; Mosser, G.; Belamie, E. Liquid crystallinity in collagen systems in vitro and in vivo. *Curr. Opin. Colloid Interface Sci.* **2008**, *13*, 303–313. [CrossRef]
26. Nyström, G.; Arcari, M.; Mezzenga, R. Confinement-induced liquid crystalline transitions in amyloid fibril cholesteric tactoids. *Nat. Nanotechnol.* **2018**, *13*, 330–336. [CrossRef] [PubMed]
27. Bagnani, M.; Nyström, G.; De Michele, C.; Mezzenga, R. Amyloid Fibrils Length Controls Shape and Structure of Nematic and Cholesteric Tactoids. *ACS Nano* **2019**, *13*, 591–600. [CrossRef] [PubMed]
28. Tam-Chang, S.W.; Huang, L. Chromonic liquid crystals: Properties and applications as functional materials. *Chem. Commun.* **2008**, *17*, 1957–1967. [CrossRef] [PubMed]
29. Zanchetta, G.; Giavazzi, F.; Nakata, M.; Buscaglia, M.; Cerbino, R.; Clark, N.A.; Bellini, T. Right-handed double-helix ultrashort DNA yields chiral nematic phases with both right-and left-handed director twist. *Proc. Natl. Acad. Sci. USA* **2010**, *107*, 17497. [CrossRef] [PubMed]
30. De Michele, C.; Zanchetta, G.; Bellini, T.; Frezza, E.; Ferrarini, A. Hierarchical Propagation of Chirality through Reversible Polymerization: The Cholesteric Phase of DNA Oligomers. *ACS Macro Lett.* **2016**, *5*, 208–212. [CrossRef]
31. Taylor, M.P.; Herzfeld, J. Liquid-crystal phases of self-assembled molecular aggregates. *J. Phys. Condens. Matter* **1993**, *5*, 2651–2678. [CrossRef]
32. Van der Schoot, P.; Cates, M.E. Growth, Static Light Scattering, and Spontaneous Ordering of Rodlike Micelles. *Langmuir* **1994**, *10*, 670–679. [CrossRef]
33. Onsager, L. The Effects of Shape on the Interaction of Colloidal Particles. *Ann. N. Y. Acad. Sci.* **1949**, *51*, 627. [CrossRef]
34. Straley, J.P. Theory of piezoelectricity in nematic liquid crystals, and of the cholesteric ordering. *Phys. Rev. A* **1976**, *14*, 1835. [CrossRef]

35. Varga, S.; Jackson, G. Study of the pitch of fluids of electrostatically chiral anisotropic molecules: Mean-field theory and simulation. *Mol. Phys.* **2006**, *104*, 3681. [CrossRef]
36. Wensink, H.H.; Jackson, G. Generalized van der Waals theory for the twist elastic modulus and helical pitch of cholesterics. *J. Chem. Phys.* **2009**, *130*, 234911. [CrossRef] [PubMed]
37. Odijk, T. Pitch of a Polymer Cholesteric. *J. Phys. Chem.* **1987**, *91*, 6060. [CrossRef]
38. Harris, A.B.; Kamien, R.D.; Lubensky, T.C. Molecular chirality and chiral parameters. *Rev. Mod. Phys.* **1999**, *71*, 1745. [CrossRef]
39. Odijk, T. Theory of Lyotropic Polymer Liquid Crystals. *Macromolecules* **1986**, *19*, 2313–2329. [CrossRef]
40. Franco-Melgar, M.; Haslam, A.J.; Jackson, G. A generalisation of the Onsager trial-function approach: Describing nematic liquid crystals with an algebraic equation of state. *Mol. Phys.* **2008**, *106*, 649. [CrossRef]
41. Wensink, H.H.; Vroege, G.J. Isotropic-nematic phase behavior of length-polydisperse hard rods. *J. Chem. Phys.* **2003**, *119*, 6868–6882. [CrossRef]
42. Goossens, W.J.A. A Molecular Theory of the Cholesteric Phase and of the Twisting Power of Optically Active Molecules in a Nematic Liqud Crystal. *Mol. Cryst. Liq. Cryst.* **1971**, *12*, 237–244. [CrossRef]
43. Varga, S.; Jackson, G. Simulation of the macroscopic pitch of a chiral nematic phase of a model chiral mesogen. *Chem. Phys. Lett.* **2003**, *377*, 6–12. [CrossRef]
44. Odijk, T. Elastic constants of nematic solutions of rod-like and semi-flexible polymers. *Liq. Cryst.* **1986**, *1*, 553–559. [CrossRef]
45. Dussi, S.; Belli, S.; van Roij, R.; Dijkstra, M. Cholesterics of colloidal helices: Predicting the macroscopic pitch from the particle shape and thermodynamic state. *J. Chem. Phys.* **2015**, *142*, 074905. [CrossRef] [PubMed]
46. Kolli, H.B.; Frezza, E.; Cinacchi, G.; Ferrarini, A.; Giacometti, A.; Hudson, T.S.; De Michele, C.; Sciortino, F. Self-assembly of hard helices: A rich and unconventional polymorphism. *Soft Matter* **2014**, *10*, 8171–8187. [CrossRef] [PubMed]
47. Wensink, H.H.; Anda, L.M. Chiral assembly of weakly curled hard rods: Effect of steric chirality and polarity. *J. Chem. Phys.* **2015**, *143*, 144907. [CrossRef] [PubMed]
48. Tombolato, F.; Ferrarini, A.; Grelet, E. Chiral Nematic Phase of Suspensions of Rodlike Viruses: Left-Handed Phase Helicity from a Right-Handed Molecular Helix. *Phys. Rev. Lett.* **2006**, *96*, 258302. [CrossRef] [PubMed]
49. Šolc, K. Cloud-Point Curves of Polymers with Logarithmic-Normal Distribution of Molecular Weight. *Macromolecules* **1975**, *8*, 819–827. [CrossRef]
50. Kiss, L.B.; Söderlund, J.; Niklasson, G.A.; Granqvist, C.G. New approach to the origin of lognormal size distributions of nanoparticles. *Nanotechnology* **1999**, *10*, 25. [CrossRef]
51. Schulz, G.V. Über die kinetik der kettenpolymerisationen. *Z. Physik. Chem. B* **1939**, *43*, 25–46. [CrossRef]
52. Zimm, B.H. Apparatus and Methods for Measurement and Interpretation of the Angular Variation of Light Scattering; Preliminary Results on Polystyrene Solutions. *J. Chem. Phys.* **1948**, *16*, 1099–1116. [CrossRef]
53. Speranza, A.; Sollich, P. Isotropic-nematic phase equilibria of polydisperse hard rods: The effect of fat tails in the length distribution. *J. Chem. Phys.* **2003**, *118*, 5213–5223. [CrossRef]
54. Ferreiro-Córdova, C.; Wensink, H.H. Spinodal instabilities in polydisperse lyotropic nematics. *J. Chem. Phys.* **2016**, *145*, 244904. [CrossRef] [PubMed]

© 2019 by the authors. Licensee MDPI, Basel, Switzerland. This article is an open access article distributed under the terms and conditions of the Creative Commons Attribution (CC BY) license (http://creativecommons.org/licenses/by/4.0/).

Review

Liquid-Crystalline Dispersions of Double-Stranded DNA

Yuri Yevdokimov [1,*,†], Sergey Skuridin [1,†], Viktor Salyanov [1,†], Sergey Semenov [2] and Efim Kats [3]

1. Engelhardt Institute of Molecular Biology of the Russian Academy of Sciences, Vavilova, St. 32, Moscow 119991, Russia; lancet-bio@bk.ru (S.S.); vsalyanov@yandex.ru (V.S.)
2. National Research Centre "Kurchatov Institute", Kurchatova Sq. 1, Moscow 123182, Russia; semenov_sv@nrcki.ru
3. Landau Institute for Theoretical Physics of the Russian Academy of Sciences, Academician Semenov Ave. 1a, Chernogolovka, Moscow Region 142432, Russia; kats@landau.ac.ru
* Correspondence: yevdokim@eimb.ru; Tel.: +7-499-135-97-20; Fax: +7-499-135-14-05
† These authors contributed equally to this manuscript.

Received: 13 February 2019; Accepted: 17 March 2019; Published: 20 March 2019

Abstract: In this review, we compare the circular dichroism (CD) spectra of liquid-crystalline dispersion (LCD) particles formed in PEG-containing aqueous-salt solutions with the purpose of determining the packing of ds DNA molecules in these particles. Depending on the osmotic pressure of the solution, the phase exclusion of ds DNA molecules at room temperature results in the formation of LCD particles with the cholesteric or the hexagonal packing of molecules. The heating of dispersion particles with the hexagonal packing of the ds DNA molecules results in a new phase transition, accompanied by an appearance of a new optically active phase of ds DNA molecules. Our results are rationalized by way of a concept of orientationally ordered "quasinematic" layers formed by ds DNA molecules, with a parallel alignment in the hexagonal structure. These layers can adopt a twisted configuration with a temperature increase; and as a result of this process, a new, helicoidal structure of dispersion particle is formed (termed as the "re-entrant" cholesteric phase). To prove the cholesteric pattern of ds DNA molecules in this phase, the "liquid-like" state of the dispersion particles was transformed into its "rigid" counterpart.

Keywords: liquid-crystalline dispersions of DNA; cholesteric and hexagonal packing of DNA; theoretically calculated and experimental circular dichroism spectra; textures of the DNA liquid-crystalline phases; "re-entrant" cholesteric phase of DNA; anthracyclines drugs; chelate complexes; nanobridges; "rigid" particles of DNA

---

1. Introduction:

The packing of double-stranded (ds) DNA molecules in liquid-crystalline (LC) phases is still the focus of many experimental and theoretical works.

Since 1961 it has been known that the dissolution of lyophilized ds DNA samples (of high molecular mass) in aqueous-salt solutions and the adjustment of concentration with the buffer solution is accompanied by the assembling of the adjacent molecules and the formation of condensed phases. Historically, the first of such phases were cholesteric [1] and columnar hexagonal phases [2–10].

An alternative method to prepare the ds DNA phases is to increase ds DNA concentration by ultrafiltration through a membrane the pore size of which allows the passage of water and ions but not ds DNA molecules [11–14].

One can note that most previous work and reviews are dedicated to the problem of the packing of ds DNA molecules in the formed phases [8,15–18]. The patterns of the ds DNA molecules packing in different phases have been investigated by electron (and cryo-electron) microscopy as well as the analysis of thin layer textures of these phases by polarizing microscopy [19–22].

The X-ray diffraction analysis was used to measure the ds DNA interhelix spacing in the formed phases. According to these studies, the adjacent ds DNA molecules are ordered in the formed phases at distances of 2.5–5.0 nm, that is they acquire the properties of a *crystal* and according to the results of the analysis of their thin layer textures the DNA molecules are mobile, that is they retain the properties of a *liquid* (see, for instance, [23–25]).

As a result of all of these studies the term "ds DNA liquid-crystalline phases" is proposed and it is possible to establish the details of the ds DNA packing in the formed LC phases. The cascade of DNA phases established in these works looks as follows: isotropic phase → (blue phase?) → cholesteric phase → columnar phase → crystalline phase [11,16–18,25,26].

Bulk ds DNA LC phases represent, by themselves, viscous solutions in which adjacent DNA molecules with phosphate groups neutralized by cations are ordered while keeping their ability to slide relative to each other.

Hence, ds DNA molecules undergo a spontaneous transition from an isotropic to a compact (condensed) state at the increase of their concentration in aqueous-salt solutions [25,27]. The latter state in this transition is, as a rule, characterized by the hexagonal packing [16,21,28,29] enabling a high packing density of adjacent ds DNA molecules ("local" order in the arrangement of the centres of mass of the molecules detected by small-angle X-ray scattering).

The mean interhelix distance between ds DNA molecules in the cholesteric phase varies from about 5.0 mm to 3.2 nm and in the case of the hexagonal phase from 3.2 nm to 2.3 nm [25].

Besides, it has been shown that the mixing of the ds DNA molecules and aqueous-salt solutions of some strongly hydrophilic polymers poly (ethylene glycol), poly (acrylic acid) or poly (vinyl pyrrolidone) leads to the phase exclusion of the DNA (Figure 1). This process has been known since Lerman's experiments (1971) as "ψ-condensation" (*psi* is the acronym for *polymer-salt-induced*) [30–33]). This method of the phase exclusion, depending on two parameters, that is, ds DNA molecular mass and its concentration, can be realized as the intramolecular or intermolecular assembling of these molecules.

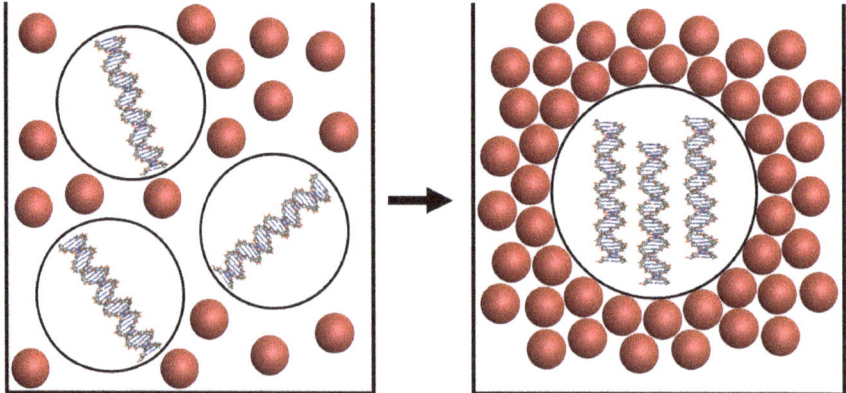

Figure 1. Simplest scheme of the phase exclusion of low molecular mass, linear, rigid ds DNA molecules from an aqueous-salt PEG-containing solution. (**Left panel**)—At concentration of PEG below the "critical" value, there is enough a "free space" for ds DNA molecules. (**Right panel**)—At concentration of PEG higher than the "critical" value, the ordered assembling of adjacent ds DNA molecules takes place. ●—Image of PEG molecules.

In the first case, the phase exclusion of the extremely high molecular mass ds DNA (molecular mass $\geq 10 \times 10^6$ Da) performed at a very low DNA concentration, leads to the formation of toroid-like particles consisting of a single DNA molecule [30,34]. The toroid-like DNA particles have been detected experimentally under various conditions [35,36].

Despite of the fact that there are current theoretical attempts [37,38] to describe this pattern of DNA condensation, a universal theory that explains the formation of DNA toroidal condensates does not exist.

In the second case, the phase exclusion (condensation) of the low molecular mass, rigid, linear ds DNA (molecular mass $< 1 \times 10^6$ Da) from aqueous-salt poly (ethylene glycol) (PEG) solutions [39,40] or mixtures of PEG with a mineral oil [41] at room temperature is accompanied by the assembling of adjacent molecules and by the formation of DNA dispersions.

The particles of the low molecular mass ds DNA dispersions are "microscopic droplets of concentrated DNA solution", which cannot be "taken in hand" or "directly seen." A "liquid-like" pattern of packing of ds DNA molecules in the particles prevents their immobilization on the surface of a membrane filter.

Theoretical estimations of the ds DNA dispersion particle size, based on the data obtained by different methods (low-speed centrifugation, UV-light scattering, dynamic light scattering, etc.), show that for the DNA with a molecular mass of about $(0.6–0.8) \times 10^6$ Da the mean particle diameter is close to 500–1000 nm and one particle contains about 10^4 DNA molecules [42]. These results have been confirmed by the direct atomic force microscopy data on specially prepared "rigid" (gel-like) ds DNA dispersion particles, immobilized on the surface of nuclear membrane filters [43].

The ds DNA ordering in dispersion particles is interesting from several viewpoints. These particles are of biological interest, because ds DNA dispersions represent the simplest model systems for DNA packing in nature. Indeed, in vivo this state is observed in bacteriophage heads, bacterial nucleoids, dinoflagellate chromosomes and the sperm nuclei of many vertebrates (such as humans, horses, rabbits) [17,44] that are, in fact, microscopic bodies that contain densely packed DNA molecules.

It is worth stressing that the patterns of DNA molecules packing in the dispersion particles can significantly differ from the mode typical of the bulk LC phases [45,46]. Indeed, we should keep in mind that we are dealing with finite size condensed ds DNA particles, not with bulk LC systems. In such confined geometry there is a diverse variety of generally metastable cholesteric-like structures, first of all because the boundary conditions (surface anchoring) are incompatible with the optimal cholesteric twist. Multi stability of possible cholesteric structures is the result of this frustration.

Finally, due to the reactive capacity of the DNA molecules, the dispersion can provide background for biosensors in analytical kits designed for detecting biologically active and chemically important compounds.

Dispersion particles have several characteristic features. First, the polymer (PEG) is not contained in dispersion particles. Second, adjacent DNA molecules are separated by distance of 2.5–5.0 nm, that is, particles have certain crystal-like properties. Third, the concentration of DNA in dispersion particles (or packing density of DNA molecules) is very high and it depends on the osmotic pressure of a PEG-containing solution (or PEG concentration in the solution).

Clearly, that the maximal density of ds DNA molecules in the dispersion particles can be achieved at their hexagonal packing. Hexagonal packing may be schematically represented, as shown in Figure 2. Ordering in a single direction in the case of hexagonal packing is evident from Figure 2. However, this packing does not correspond to the structure of a true crystal, because there are disordered water molecules between the ds DNA molecules. The shown structure does not possess a long-range positional order. The ds DNA molecules have some disorder around their positions; they can slide and bend in respect to each other, as well as rotate around their long axis. This corresponds to the "liquid" character of the ds DNA molecules packing. These facts allow one to describe such particles using the "liquid-crystalline dispersion (LCD) DNA particle" term.

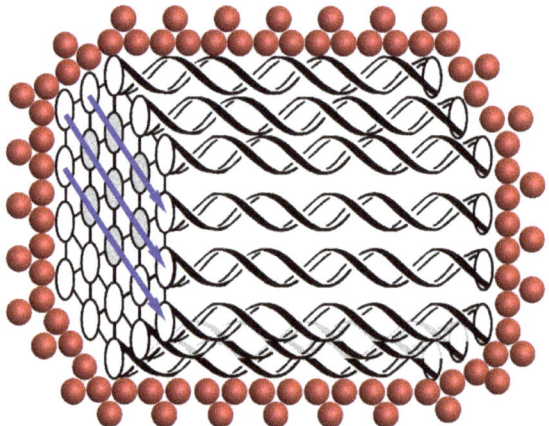

Figure 2. Hypothetical scheme of the hexagonal packing of linear ds DNA molecules in a dispersion particle formed in an aqueous-salt PEG-containing solution. ⬤—Schematic images of PEG molecules; the blue arrows indicate the quasinematic layers of DNA molecules. The three main directions of the hexagonal array permit one to define the quasinematic layers in the structure. Quasinematic layers, even with a constant distance between them, can be rotated by a small angle with a change in the osmotic pressure or temperature. It leads to the formation of a spatially twisted structure with an abnormal optical activity.

The minimization of the excluded volume of adjacent linear, rigid, ds DNA molecules leads to their parallel, unidirectional (nematic-like) alignment. Regardless of the dense packing, some authors assume that the molecules form so-called "quasinematic" layers (marked by blue arrows in Figure 2) [47–50]. Double-stranded nucleic acid molecules lie in the plane of the layers, with the layer thickness being close to the intermolecular distance [25].

It should be noted that the packing pattern of the ds DNA molecules in bulk LC phases can be determined by various methods, while additional studies are required to determine, how the DNA molecules are packed in dispersion particles obtained as a result of their phase exclusion from aqueous-salt solutions of polymers (for instance, PEG-containing solutions).

2. Some Peculiarities of the CD Spectra of ds DNA Liquid-Crystalline Dispersions

Very useful and robust global information about the secondary structure of ds DNA molecules and their pattern of packing within spatially distributed, independent, small size dispersion particles can be obtained by application of the circular dichroism method [51,52].

In Reference [52] a phenomenological approach based on the theory [53–57] of electromagnetic wave absorption by large molecular aggregates has been developed. This approach permits the analysis of many peculiarities of the circular dichroism (CD) spectra of ds DNA LCD particles.

Let us briefly comment on this approach. When the size of a molecular aggregate is comparable to light wavelength, its optical properties depend on both the size and the shape of dispersion particles. The properties are determined by long-distance interactions between chromophores that occur in a given structure. To calculate optical characteristics of the structure, a particle can be divided into cells and each cell is considered as a single effective chromophore. These effective chromophores, which are absorbing dipoles, have an ordering that is similar (but not identical, see the note of caution above) to a cholesteric helix with the pitch P.

The approach proceeds from the simplifying assumption that the DNA particles ("molecular aggregates") are cubic in shape. Thus, a single particle is modelled as a set of quasinematic layers that are the distance d apart and contain absorbing dipoles with the rotation angle $\Delta \varphi$ between two adjacent layers. Accordingly, $P = 2\pi d/\Delta\varphi$. The electric field at the point where a particular chromophore occurs is a superposition of the field of incident light and the electric fields that are generated by induced dipoles of all the other chromophores of the system. As a result, light absorption is a process that involves the whole molecular aggregate and the electric field within a particle is obtained by solving a self-consistent set of linear equations for the field amplitudes at each effective chromophore.

The theoretical studies predict an appearance of an intense band in the CD spectrum located in the region of absorption of the original chromophores that are contained in the molecular aggregate (condensed phase) with the *cholesteric pattern* of molecules packing [58–61]. Its value is related mainly to the spatial parameters of particles (their diameter D) and the magnitude of the cholesteric pitch P.

The sign of the helical twist of the spatial structure of the molecular aggregate is determined by the sign of an intense band in the CD spectrum (i.e., a positive band corresponds to a right-handed helical arrangement and a negative band means a left-handed twist of adjacent quasinematic layers of molecules).

The theory's assumptions are applicable both to the original chromophores of the system and the "external" (additional) chromophores introduced into the structure of the molecular aggregates with a cholesteric packing of molecules. If the molecules of the "external" chromophores are arranged in the structure of the cholesteric molecular aggregate specifically, the theories [58–61] predict the appearance of two intense bands in different regions of the CD spectrum. Random arrangement of any type of chromophores in the molecular aggregate is unaccompanied by any considerable change in the amplitude of their CD bands.

After Norden paper [62] an intense band in the CD spectrum is expressed as ΔA (in optical units) and reflects a so-called "structural circular dichroism". In order to stress the difference between the "molecular circular dichroism" and the "structural circular dichroism", the term "abnormal band" has been used to designate an intense band in the CD spectrum [63].

Hence, according to theoretical considerations, an abnormal band in the CD spectrum in the absorption region of the original chromophores in the content of a cholesteric molecular aggregate can be considered direct evidence of its helically twisted spatial structure.

This approach was applied to the calculation of the CD spectra of ds DNA LCD particles and the dependence upon their structural parameters [52,63,64].

The particles of ds DNA dispersions are considered polycrystalline objects with random distribution and orientation of individual particles, possessing their own absorption in the UV-region of the spectrum due to the presence of the chromophores (nitrogen bases) in the content of DNA molecules. The theory takes into account both the quasinematic structure of the ds DNA molecules packing (Figure 2) and the data shown above, concerning these dispersion particles. In particular, experimental data enable us to treat the particles of dispersions, formed by linear, rigid, ds DNA molecules of low molecular mass (about 10^6 Da), as spheres of diameter D. For these particles, because of the inherent rigidity of these molecules, the LC ordering is specific at the phase exclusion from PEG-containing solutions.

The distance between ds DNA molecules is determined by a balance between the repulsive intermolecular forces and the compressing osmotic pressure of the PEG solution. Besides, ds DNA molecules possess several levels of chirality (helical secondary structure of ds DNA molecules, helical distribution of counter-ions near the DNA surface as well as asymmetry of C-atoms in sugar residues). The chiral and anisotropic properties of ds DNA molecules favour the helical twisting of adjacent molecules.

Note that if molecules in the content of a dispersion particle rotate independently around their long axes, their chiral interactions can be averaged and smoothed-out [65,66]. The non-zero chiral interactions require rotational biaxial correlations between ds DNA molecules and the behaviour of adjacent ds DNA molecules in dispersion particles should be spatially correlated to make the chiral interaction pronounced [66]. At room temperature, this correlation is possible only in the case of large enough distances between ds DNA molecules in quasinematic layers. In this case, the unidirectional alignment of ds DNA molecules in quasinematic layers of dispersion particles competes with the tendency of these molecules to form a spatially twisted (cholesteric) structure. The spatial correlation is very limited in the case of dense packing of ds DNA molecules.

Hence, we consider low molecular mass ds DNA molecules ordered in particles of dispersions as continuous cholesteric LC phase.

In such cases, the anisotropy of the absorption of nitrogen bases (chromophores) for the linear polarizations of the light must show itself as an intense band in the CD spectrum.

The sign of the band in the CD spectrum in the region of the absorption of nitrogen bases must depend on the orientation of the planes of these components in respect to the long axis of the DNA molecule.

The experimentally measured CD spectrum of the initial, linear, ds DNA molecules in aqueous-salt solution (isotropic state of DNA) is well-known [67] and is shown in Figure 3a.

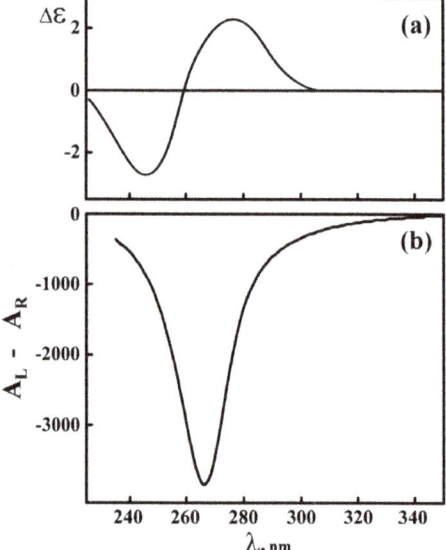

Figure 3. The CD spectrum of an aqueous-salt (0.3 M NaCl + 0.002 M Na$^+$-phosphate buffer) solution of linear double-stranded B-form of DNA (**a**) and the theoretically calculated CD spectrum for ds DNA CLCD (**b**). C_{DNA} = 10 µg·mL^{-1}, P = 2500 nm, D = 500 nm. $\Delta A = (A_L - A_R) \times 10^{-6}$ optical units, L = 1 cm.

The ds DNA molecules possess the parameters of a classical B-form; this is confirmed by the results of small-angle X-ray scattering [68,69] and by measured $\Delta \varepsilon$ value. The *"molecular circular dichroism"* (expressed as $\Delta \varepsilon$ value, $\Delta \varepsilon_{max}$ ~ 2.5 M^{-1}·cm^{-1}), that is, the physical constant, is usually used to describe the peculiarities of *isolated* nitrogen bases or *individual* DNA molecules. Its value can be calculated theoretically [67].

The theoretically calculated CD spectrum of a DNA dispersion formed in a PEG-containing solution is given in the Figure 3b. One can see that, in accordance with the theoretical assumption given above, the formation of a dispersion is followed by an appearance of an intense (abnormal) negative band in the CD spectrum located in the region of the nitrogen bases absorption. An appearance of this band in the CD spectrum of the ds DNA LCD means that the purine and pyrimidine nitrogen bases do play the role of "chromophores," providing information about the spatial packing of DNA molecules in particles of liquid-crystalline dispersions. This conclusion corresponds to the intuitive expectation that the cholesteric packing is a specific property of any molecule having geometrical and optical anisotropy and chirality. Hence, rigid, anisotropic, double-stranded nucleic acid molecules (DNA, RNA, etc.) tend to realize their potential tendency to the cholesteric mode of packing in dispersion particles.

However, the condensation of DNA molecules is not by itself a sufficient condition for the appearance of an intense band in the CD spectrum, since many aggregated forms of DNA (for instance, the aggregates formed by single-stranded DNA molecules) fail to show intense band in their CD spectra.

A comparison of Figure 3b to Figure 3a allows one to conclude that the intense negative band in the CD spectrum reflects a so-called "*structural circular dichroism*" and must be expressed as ΔA (in optical units).

The theory [52] does predict the following features of the CD spectrum of ds DNA dispersion particles.

(1) The formed particles are "coloured" because DNA molecules contain chromophores (nitrogen bases) that absorb UV irradiation.

In this case, theory predicts an appearance of an abnormal band in the CD spectrum in the region of absorption of the ds DNA nitrogen bases. This is the univocal indication of the macroscopic (cholesteric) twist in the orientation of quasinematic layers formed by ds DNA molecules in dispersion particles. The term "cholesteric liquid-crystalline dispersions" (CLCDs) is used to signify these dispersions [64].

(2) The theoretically calculated amplitude of the abnormal band in the CD spectrum increases as the diameter of the ds DNA CLCD particles grows (Figure 4). However, if the diameter of the DNA CLCD particles reaches the minimal value about of 50 nm, the amplitude of the abnormal band in the CD spectrum decreases so sharply that it no longer can be distinguished from that in the CD spectrum typical of the initial linear DNA. This result indicates that in the case of the formation of CLCDs with a particle diameter of 50 nm their presence cannot be registered by CD spectroscopy.

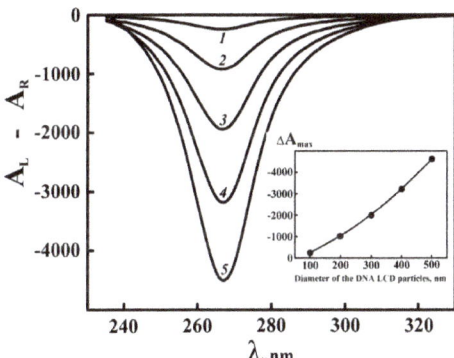

Figure 4. Theoretically calculated CD spectra of ds DNA CLCDs with different size (*D*) of particles: D-100, 200, 300, 400 and 500 nm (curves 1–5). C_{DNA} = 10 µg·mL^{-1}, P = 2500 nm; $\Delta A = (A_L - A_R) \times 10^{-6}$ optical units, L = 1 cm. Inset: The dependence of the amplitude (ΔA_{max}) the negative band in the CD spectra of the DNA CLCD on size (*D*) of particles.

(3) The smaller the pitch (*P*)value of the DNA cholesteric structure (i.e., the greater the twist—the angle between the adjacent quasinematic layers of DNA molecules in the helical structure of the cholesteric), the more intense the band in the CD spectrum (Figure 5).

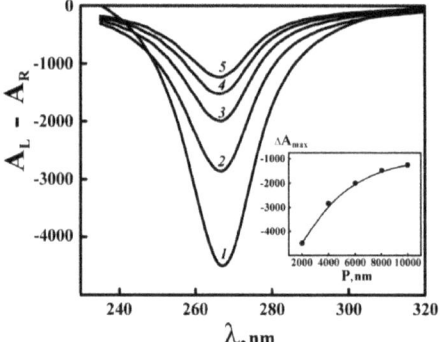

Figure 5. Theoretically calculated CD spectra of ds DNA CLCDs whose particles have different cholesteric pitch (P). Curves 1–5: P-2000, 4000, 6000, 8000 and 10,000 nm. $C_{DNA} = 10\ \mu g \cdot mL^{-1}$, $D = 500$ nm. Inset: The dependence of the amplitude (ΔA_{max}) of the negative band in CD spectra on pitch (P) of the spatially twisted structure of the DNA CLCD.

Conversely, the more untwisted the cholesteric structure, the lower the amplitude of the band in the CD spectrum of the CLCD. The calculations have shown that at a P value of about 30 m and constant structural properties of DNA molecules, the amplitude of the abnormal band in the CD spectrum is quite close to the amplitude of the band characteristic of isolated linear DNA molecules.

(4) The sign of the abnormal band in the CD spectrum of DNA CLCD depends on the sense of the twist of the spatial cholesteric structure formed by DNA molecules packed in CLCD particles (Figure 6). The negative sign of the abnormal band in the CD spectrum proves the left-handed cholesteric twist of quasinematic layers formed by the right-handed DNA molecules (B-form) in particles of dispersions.

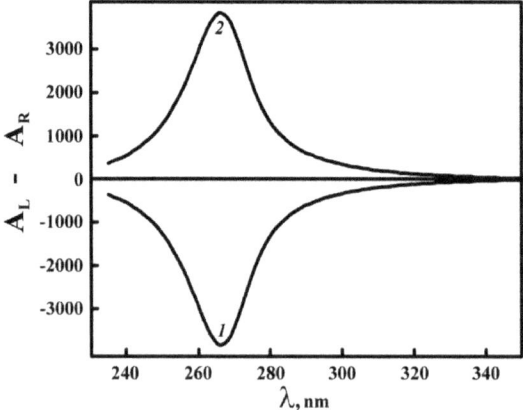

Figure 6. Theoretically calculated CD spectra of ds DNA CLCDs with left- (curve 1) and right-handed (curve 2) twist of spatial structure. $C_{DNA} = 10\ \mu g \cdot mL^{-1}$, $D = 500$ nm, $P = 2500$ nm. $\Delta A = (A_L - A_R) \times 10^{-6}$ optical units, $L = 1$ cm.

(5) Very small alterations in the base pair sequence or in parameters of the secondary structure of nucleic acid molecules can be sufficient to cause the change from the left-handed to the right-handed twist of the quasinematic layers in particles of CLCD. This means, that there are two types (left-handed and right-handed) of cholesteric structures formed by ds nucleic acids.

(6) The "external" (additional) chromophores with the absorption bands that do not coincide with the nitrogen bases absorption band can be chemically incorporated into the structure of the ds DNA molecules.

Such a situation is possible at the insertion (intercalation) of coloured biologically active compounds, for instance, antibiotic of the anthracycline group—daunomycin (DAU)—between DNA base pairs.

Since ds DNA molecules are ordered in quasinematic layers and "external" chromophores (in our case DAU) are rigidly fixed in respect to the long axis of DNA molecules, the "external" chromophores are automatically included in the composition of quasinematic layer.

For this case, the theory predicts an appearance of two bands located in different regions of the CD spectrum. One of the bands is located in the region of DNA chromophore (nitrogen bases) absorption ($\lambda \sim 270$ nm), the other one is situated in the region of DAU absorption ($\lambda \sim 510$ nm) (Figure 7). Both of the bands have negative signs and their amplitudes are much larger in comparison to that typical of the molecular circular dichroism of nitrogen bases and DAU chromophores. The coincidence of the signs of these bands shows that the DAU molecules are located in respect to the DNA molecules long axis the same way as the nitrogen base pairs, that is, the angle between the "external" chromophore molecules plane and the DNA molecules long axis is close to 90°. Such coincidence of the signs of two abnormal bands in different regions of the CD spectrum is only possible if DAU molecules intercalate between DNA nitrogen base pairs.

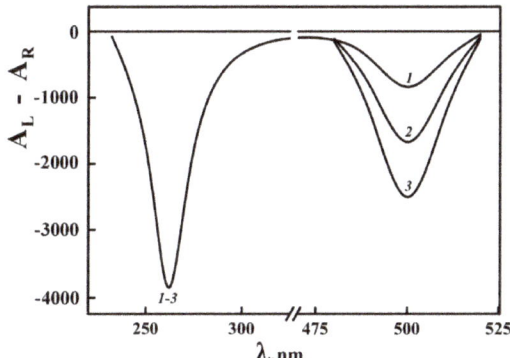

Figure 7. Theoretically calculated CD spectra of ds DNA CLCD treated with daunomycin (curves 1–3 correspond to different DAU concentrations; 3 > 1). $C_{DNA} = 10$ µg·mL^{-1}, $D = 500$ nm, $P = 2500$ nm. $\Delta A = (A_L - A_R) \times 10^{-6}$ optical units, L = 1 cm.

Figure 7 shows that the amplitude of the negative band in the region of DAU absorption increases as the extent of DAU molecules intercalation between the DNA nitrogen base pairs increases, reaching the equilibrium state.

Hence, the appearance of two independent bands in different regions of the CD spectrum is definite evidence in favour of the cholesteric pattern of packing of the ds DNA molecules in the CLCD particles.

3. Formation of LCDs as a Result of Phase Exclusion of ds DNA Molecules from PEG Solutions and Properties of these Dispersions at Room Temperature

Double-stranded DNA LCDs have been formed as the result of a phase exclusion when aqueous-salt solutions of DNA were mixed with aqueous-salt solutions of a synthetic chemically neutral polymer—poly (ethylene glycol) (PEG) [27,39].

The phase exclusion of linear ds DNA molecules with the "standard" molecular mass from aqueous-salt PEG solutions [70] was accompanied by the formation of dispersions, which possesses an abnormal band in the CD spectrum (Figure 8a, curves 1–4).

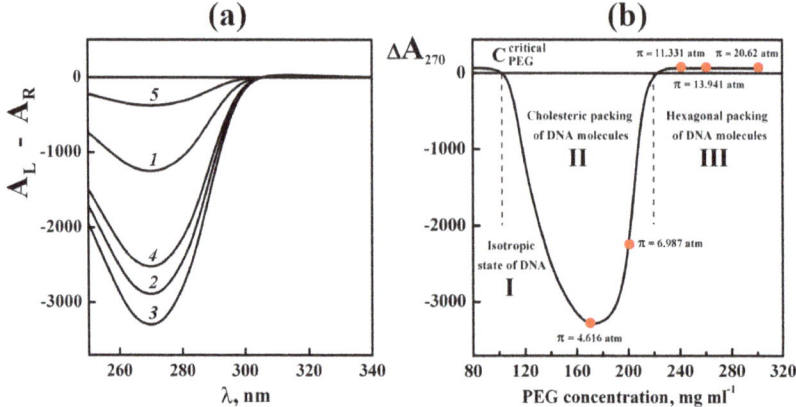

Figure 8. (a) The CD spectra of the DNA LCDs formed in aqueous-salt solutions with different PEG concentration. 1—C_{PEG} = 120 mg·mL^{-1} (π = 2.062 atm), 2—C_{PEG} = 150 mg·mL^{-1} (π = 3.422 atm), 3—C_{PEG} = 170 mg·mL^{-1} (π = 4.616 atm), 4—C_{PEG} = 200 mg·mL^{-1} (π = 6.987 atm), 5—C_{PEG} = 210 mg·mL^{-1} (π = 7.839 atm). C_{DNA} = 10 µg·mL^{-1}, 0.3 M NaCl + 0.002 M Na$^+$-phosphate buffer. $\Delta A = (A_L - A_R) \cdot 10^{-6}$ optical units, T = 22 °C, L = 1 cm. (b) The generalized dependence of the amplitude of the band in the CD spectra of DNA LCDs (λ = 270 nm) upon PEG concentration. C_{DNA} = 10 µg·mL^{-1}, 0.3 M NaCl + 0.002 M Na$^+$-phosphate buffer. $\Delta A_{270} \times 10^{-6}$ optical units, T = 22 °C, L = 1 cm. $C_{PEG}^{critical}$—"critical" concentration of PEG at which the formation of DNA LCDs begins. p—osmotic pressure of PEG solution. domain I—isotropic ds DNA solution; domain II and domain III—the existence of the ds DNA LCDs with cholesteric and hexagonal packing of molecules in particles, respectively.

The appearance of the abnormal negative band in the CD spectrum located in absorption region of ds DNA nitrogen bases (λ270 nm) is the direct evidence of their introduction (as chomophores) into the helically twisted structure of dispersion particles [58,59].

Taking into account the rigid position of nitrogen bases in respect to the long axes of the ds DNA molecules, one can infer, in turn, the existence of a helically twisted spatial (cholesteric) structure of quasinematic layers in dispersion particles [70]. (Under the conditions used, the single-stranded (or denatured) DNA does not condense and cannot go into the regular composition of dispersion particles [71]).

The decrease in the amplitude of the abnormal band in the CD spectrum (to zero level) with an increase in the PEG concentration (Figure 8b, domain **III**) is indicative of the untwisting of the spatial helical structure of dispersion particles [72].

It is worth reminding that in the range of osmotic PEG pressures up to ~20 atm (Figure 8b), DNA molecules belong to the B-form, as confirmed by the X-ray diffraction data [68,69].

Besides, the small-angle X-ray scattering data of phases that were formed as a result of the low-speed sedimentation of ds DNA dispersion particles under different PEG concentrations demonstrate the dense packing of the ds DNA molecules in dispersions formed in domains **II** and **III**.

In domain **II** (120 mg·mL^{-1} ≤ C_{PEG} ≤ 220 mg·mL^{-1}) the mean distance (d) between ds DNA molecules varies from 3.8 nm to 2.8 nm; whereas in the domain **III** (220 mg·mL^{-1} ≤ C_{PEG} ≤ 320 mg·mL^{-1}), the d value only slightly decreases (from 2.8 to 2.4 nm) [70].

Therefore, in domain **II** the LCD particles formed by ds DNA molecules at room temperature possess a structure not only with a dense packing of the linear adjacent ds DNA molecules but also with a spatially twisted arrangement of these molecules.

This conclusion is confirmed by the presence of a texture of the phase obtained at the low- speed sedimentation of dispersion particles (Figure 9). It is a "fingerprint" texture, which istypical of ds DNA cholesterics [19,25].

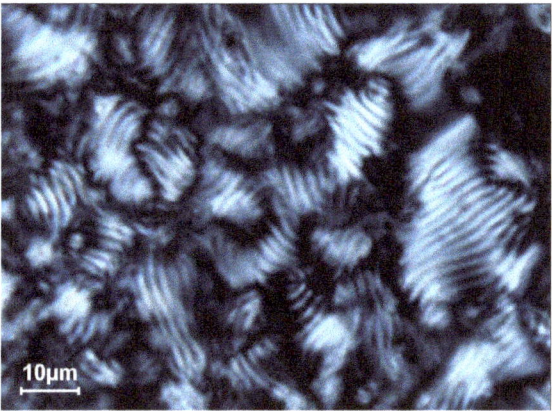

Figure 9. Fingerprint texture of cholesteric phaseobtained as the result of the low-speed sedimentation of the ds DNA particles formed in a PEG-containing aqueous-salt solution (C_{PEG} = 170 mg·mL^{-1}). Bar corresponds to 10 μm.

The ds DNA molecules spatial ordering in the LCD particles in domain **III** is significantly different from that in domain **II**. At osmotic pressure above 10 atm (PEG concentrations ~ 240–300 mg·mL^{-1}) the packing pattern can be described as unidirectional hexagonal alignment of ds DNA molecules or as a hexagonal packing with the parallel (nematic-like) alignment of orientationally ordered ds DNA quasinematic layers.

The hexagonal packing of ds DNA molecules does not possess an abnormal optical activity [25]. This statement is confirmed by the optical texture of the phase obtained as the result of the low-speed sedimentation of dispersion particles (Figure 10).

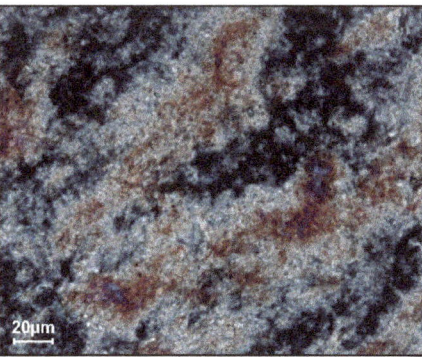

Figure 10. Thin-layer optical texture of the phase obtained as the result of the low-speed sedimentation of the ds DNA particles formed in a PEG-containing aqueous-salt solution (C_{PEG} = 240 mg·mL^{-1}). Bar corresponds to 20 μm.

This texture differs from the fingerprint texture of ds DNA cholesteric shown above. Under conditions, which correspond to the hexagonal packing of ds DNA molecules, the main part of this texture only has the attributes of an anisotropic object, where deep red and blue colors are present but does not contain any specific peculiarities.

Hence, Figure 8 shows that the phase exclusion of ds DNA molecules at room temperature leads to the following sequence of phase transitions: isotropic (I) → cholesteric (II) → hexagonal state (III) governed by an increase of the PEG concentration in a solution (PEG osmotic pressure). Two types of the condensed states (i.e., cholesteric and hexagonal) differ by the packing pattern of ds DNA molecules.

4. The Effect of Temperature on the Optical Properties of ds DNA LCDs

Figure 11 shows, as an example, the change in the CD spectrum of ds DNA CLCD formed in domain **II** (curve 1, C_{PEG} = 170 mg·mL^{-1}, π = 4.62 atm) upon heating and subsequent cooling (curves 1, 2 and 3, respectively).

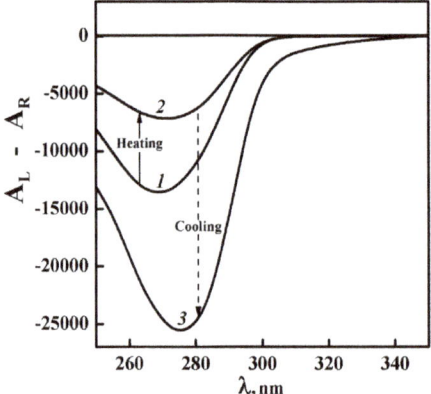

Figure 11. The CD spectra of the ds DNA CLCD formed in domain **II** (C_{PEG} = 170 mg·mL^{-1}, curve 1) upon heating (curve 2) and subsequent cooling (curve 3). 1–22°C, 2–80°C, 3–80 →22 °C. C_{DNA} = 30 μg·mL^{-1}, 0.3 M NaCl + 0.002 M Na$^+$-phosphate buffer. $\Delta A = (A_L - A_R) \times 10^{-6}$ optical units, L = 1 cm.

The well-known decrease in the amplitude of the abnormal CD band upon heating (compare curves 1 and 2 in Figure 11) is referred to as the "CD melting" [73]. It is the evidence of the untwisting of spatial helical structure of dispersion particles. The dispersion cooling is accompanied not only by the recovery of the abnormal band in the CD spectrum but also by an increase in its amplitude (curve 3). This effect reflects the "quality" of ds DNA molecules packing in dispersion particles realized in the process of phase exclusion. It depends, mainly, on the peculiarities of ds DNA molecules (secondary structure, molecular mass, etc.) [74] and it is quiteconsistent with the concept of the improvement of the cholesteric packing of dispersion particle structure as a result of the thermal "training" [75,76] of ds DNA CLCDs.

The aforementioned change in the CD spectrum upon heating and cooling is characteristic of all dispersions formed in domain **II** of PEG osmotic pressure.

The change in the CD spectra of ds DNA LCD formed in domain **III** (π = 12.43 atm, C_{PEG} = 250 mg·mL^{-1}) upon heating and cooling is shown, as an example, in Figure 12.

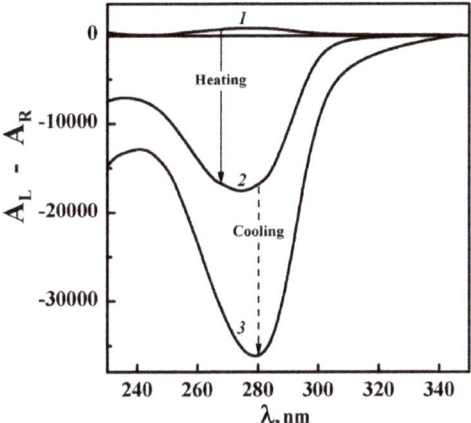

Figure 12. The CD spectra of the ds DNA LCD formed in domain **III** (C_{PEG} = 250 mg·mL^{-1}, curve 1) upon heating (curve 2) and subsequent cooling (curve 3). 1–22 °C, 2–80 °C, 3–80 → 22 °C. C_{DNA} = 30 µg·mL^{-1}, 0.3 M NaCl + 0.002 M Na$^+$-phosphate buffer. $\Delta A = (A_L - A_R) \times 10^{-6}$ optical units, L = 1 cm.

(One can see that in the initial CD spectrum of this LCD there is the band (curve 1) of unknown origin. Usually, a very low amplitude of this band is accepted as a zero value (see also Figure 8, domain **III**).

The increase in temperature is accompanied by an unusual effect: an intense negative band arises in the CD spectrum (Figure 12, curve 2) of DNA dispersion particles that did not possess such band at room temperature (curve 1).

An appearance of the intense band in domain **III** occurs at a certain "critical" temperature, which depends on the PEG osmotic pressure [77]. The higher the PEG osmotic pressure the larger the "critical" temperature value.

An intense band in the CD spectrum (Figure 12, curve 2) in the absorption region of ds DNA is related [27] to the spatial twist of nitrogen based in quasinematic layers of dispersion particles, or, more exactly, to the formation of a new phase of ds DNA molecules, which possess an abnormal optical activity. It permits us to suggest that the original hexagonal packing of ds DNA molecules in the dispersion particles changes upon heating the LCDs (obtained in domain **III**).

It is necessary to add that the optical texture of this new, optically active phase does not possess the fingerprint texture typical of classical cholesterics.

The curves characterizing the change in the abnormal bands for all ds DNA LCDs formed in domains **II** and **III** under their heating (curve 2) and cooling (curve 3) are shown in Figure 13.

One can see that the increase and the subsequent decrease in the temperature is accompanied by different optical effects, depending on the PEG osmotic pressure. For instance, at PEG osmotic pressures below 10 atm, the abnormal band increases (Figure 13, curve 3) upon cooling, whereas at PEG osmotic pressures exceeding 10 atm one observes a tendency to partial recovery of the initial abnormal optical activity of LCD. These optical effects depend on molecular mass of ds DNA molecules and on their nitrogen bases content [77].

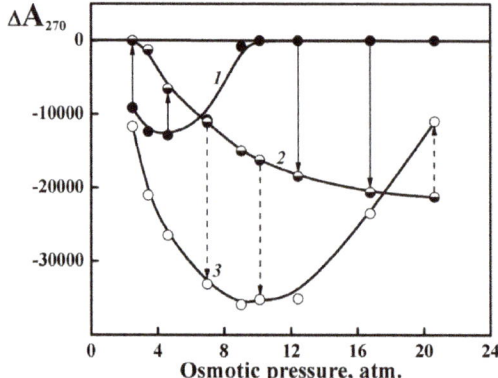

Figure 13. The dependence of an abnormal band amplitude in the CD spectra (λ = 270 nm) of the ds DNA LCDs formed in PEG-containing aqueous-salt solutions with different osmotic pressures. curve 1—room temperature, curve 2—heated to 80 °C, curve 3—cooled back from 80 °C to 22 °C. C_{DNA} = 30 µg·mL^{-1}, 0.3 M NaCl + 0.002 M Na$^+$-phosphate buffer. $\Delta A_{270} \times 10^{-6}$ optical units, L = 1 cm.

Hence, the dense hexagonal packing of ds DNA molecules in dispersion particles obtained at C_{PEG} more than 220 mg·mL^{-1} (osmotic pressure more than 10 atm) is transformed to a new optically active phase (structure) upon heating.

To proceed further we have to know, whether an optically active phase corresponds to the cholesteric structure. The matter is that this new state we found is formed at small distances between the ds DNA molecules in dispersion particles (see above) and the corresponding textures are not similar to classical cholesteric fingerprint textures.

In the next sections we present arguments supporting the cholesteric-like structure of this new optically active phase.

5. The Formation of "Rigid" Particles of the Cholesteric ds DNA Dispersions

To prove the spatially twisted (cholesteric) structure of the new optically active phase we have used well-elaborated approach [78] based on the transformation of the "liquid-like" state of ds DNA particles to the "rigid" (gel-like) state.

The main idea of this approach was formulated in the following way: "The adjacent ds DNA molecules with phosphate groups neutralized by sodium cations are resided in quasinematic layers of a single CLCD particle in the "dissolved" state and there is free space between them. The molecules of chemical substances ("guests") entering the free space due to diffusion can react with the chemical groups available of the DNA surface or even form chemical cross-links between DNA molecules in quasinematic layers of a CLCD particle. Such cross-linking can result in the formation of an "integrated" structure involving all the DNA molecules ordered in quasinematic layers of a CLCD particle. The integrated structure having a very high molecular mass will be incompatible with a PEG-containing solution. This means that the cross-linking of adjacent DNA molecules will lead to the transition of a single CLCD particle from the "liquid-like" to the "rigid" (gel-like) state."

The "rigid" ds DNA particles can be immobilized on the surface of a nuclear membrane filter and their properties can be studied by atomic force (AFM) or by electron microscopy.

This approach is similar, in fact, to gelation due to the formation of unordered, artificial chemical cross-links between adjacent molecules of polymers. However, in our case, gelation should be realized in such a way that the fixed distance between adjacent ds DNA molecules in quasinematic layers of a single CLCD particle does not change and the spatial helical structure of DNA CLCD particle is preserved. Therefore, we are dealing with a rather specific gelation process inside a single particle with nanometre-scale parameters.

As a control to the considered approach, we formed "rigid" particles by means of the cross-linking of adjacent ds DNA molecules fixed in quasinematic layers of CLCD particles. To create the cross-links (nanobridges) between ds DNA molecules we used anthracycline antibiotics.

These antibiotics are water-soluble, low molecular mass compounds, which have 4 reactive oxygen atoms in the 5, 6 and 11, 12 positions of anthracycline aglycone (Figure 14a).

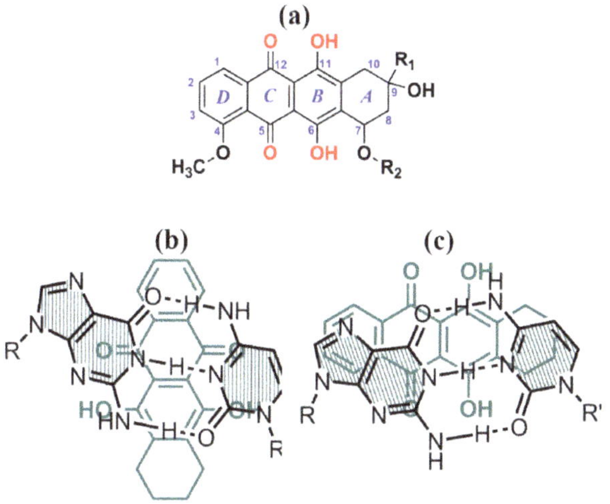

Figure 14. Panel (a) structure of aglycone part of anthracycline compounds; panel (b) and panel (c) complexes formed by DAU and ds DNA molecule; (b) scheme of intercalation complex; (c) scheme of "external" complex. View along the long axis of DNA helix.

They can form two types of complexes with ds DNA molecules, that is, the classical intercalation complex **I** (Figure 14b) at low drug concentration and nonclassical ("external") complex **II** (Figure 14c) at higher extent of binding to DNA. In the first case, an anthracycline molecule, for instance DAU, is rigidly fixed between the base pairs of DNA and synthetic polynucleotides only of the B-family (Figure 14b). Here the reactive oxygen atoms of DAU molecule become inaccessible for different chemical and electrochemical reactions even in aqueous-salt solutions but DAU molecules, being fixed in the cholesteric quasinematic layers of ds DNA, are "visible" by the CD spectroscopy. In the second case, DAU molecules form so-called "external" complexes with nucleic acids of the B- and A-families [79,80] (Figure 14c). Upon the formation of an "external" complex, the reactive groups of DAU prove to be available for chemical reactions.

The anthracyclines can form chelate complexes with bivalent metal ions like zinc, cadmium, nickel and so forth (Figure 15a). Chelate complexes produced by bivalent copper ions are of particular interest. The interest is due to the fact that the resulting chelate complexes have planar structures on account of the electronic structure of the bivalent copper ion [81–83] and the spatial structure of the anthracycline aglycons. In the case of appropriate bidentate ligands, complexation may result in the formation of a flat polymeric chain comprising up to 10 sequentially arranged molecules cross-linked by bivalent copper ions [84,85].

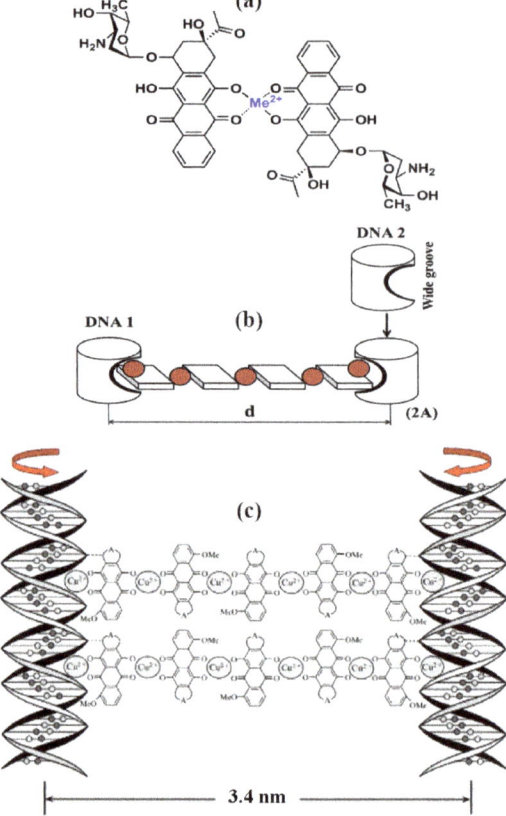

Figure 15. Panel (**a**) structure of chelate complex formed by various bivalent metal ions with oxygen groups of anthracycline compounds: panel (**b**) structure of extended chelate complex ("polymeric" chelate complex) formed between two model compounds, which possess the similar sites for complex "immersion"; panel (**c**) show a more detailed structure of nanobridges that link ds DNA molecules ordered in one quasinematic layer.

Figure 15b shows, as an example, the scheme of a "polymeric" chelate cross-link formed between two artificial molecules. The application of this scheme to ds DNA case permits to formulate a few practical requirements.

(1) In order to form polymeric cross-links between adjacent ds DNA molecules in a quasinematic layer, one can use the chemical groups of compounds deliberately "immersed" in the DNA wide groove and spatially fixed here.

(2) The "beginning" and the "end" of such cross-links must be disposed at adjacent ds DNA molecules, that is, they should link these molecules. Taking into account the mean distance between ds DNA molecules the extended cross-links formed between molecules can be named "nanobridges."

(3) Due to the spatial helical structure of ds DNA molecules (shown in Figure 15b as 1 and 2), in order to form cross-links with a fixed symmetry between the same groups of molecules "immersed" into wide grooves of ds DNA molecules it is necessary to turn DNA molecule **2** around its long axis on 180° in respect to molecule **1** (Figure 15b). It is means that cross-linking can ne only achieved at certain particular arrangement of two adjacent ds DNA molecules in quasinematic layers of CLCD particles, that is, they must be sterically "phased." Hence, spatial adjustment of the position of ds

DNA molecules within quasinematic layers of dispersion particles must be spontaneously realized to cross-link these molecules.

(4) For the sterical phasing of two parallel ds DNA molecules a certain distance between these molecules in a quasinematic layer, and, hence, definite degree of diffusion freedom for both ds DNA molecules, which is enough for a rotation of molecules around their long axes has to exist. Very close packing of ds DNA molecules in quasinematic layers will restrict the sterical phasing and formation of extended cross-links (nanobridges).

(5) Since the direction of the long axis of a polymeric chelate bridge proves to be perpendicular to the direction of the long axis of ds DNA molecules (Figure 15b), we should expect the emergence of an additional abnormal band in the CD spectrum in the absorption region of chromophores in the content of the chelate bridge.

(6) The use of ds DNA CLCD particles in this approach is of special interest because of two circumstances: (i) only in this case can one regulate the mean distance between ds DNA molecules in quasinematic layers of dispersion particles from 4.0 to 2.8 nm by changing the osmotic pressure of a PEG solution even at room temperature; (ii) only in this case the abnormal band in the CD spectra allows one to monitor the minor changes in the spatial structure of these particles and estimate the position of nanobridges in quasinematic layers.

Hence, the formation of nanobridges between adjacent ds DNA molecules is a very delicate stereochemical process, which could be realized in the case of dispersion particles that only possess the cholesteric spatial structure. It means that the considered approach can be used for the estimation of spatial structure of a new optical active phase, which arises at the heating of ds DNA dispersion particles with hexagonal packing of molecules.

The "rigid" ds DNA particles were made by means of the scheme illustrated in Figure 16.

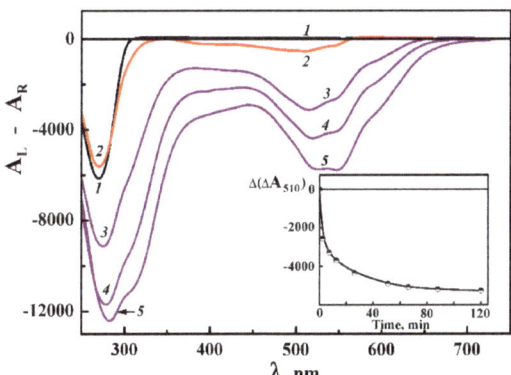

Figure 16. The CD spectra of the ds DNA CLCD formed in an aqueous-salt PEG-containing solution (curve 1) and successively processing by DAU (curve 2) and $CuCl_2$ (curves 3–5) solutions. Curve 1—the ds DNA CLCD after thermal "training." Curve 2—the ds DNA CLCD treated by DAU (C_{DAU} = 5.99 × 10^{-5} M). Curves 3–5—the ds DNA CLCD treated by DAU and $CuCl_2$ (C_{Cu} = 2.9 × 10^{-5} M) taken after 2.3 min, 12.5 min and 51 min. C_{PEG} = 170 mg·mL^{-1} (domain II), C_{DNA} = 10 µg·mL^{-1}, 0.3 M NaCl + 0.002 M Na$^+$-phosphate buffer. ($A_L - A_R$) × 10^{-6} optical units, 22 °C, L = 1 cm. Inset—the dependence of the amplitude of the band at 510 nm in the CD spectrum of ds DNA CLCD treated by DAU and $CuCl_2$ upon time.

This Figure shows the CD spectrum of ds DNA CLCD obtained as the result of the phase exclusion from PEG-containing solution at room temperature with subsequent thermal "training" (curve1). One can see that the phase exclusion of ds DNA molecules is accompanied by an appearance of an abnormal negative band in the nitrogen bases absorption region (λ 270 nm). According to the theory [51,52] an appearance of this band demonstrates the formation of ds DNA dispersion particles, which possess the cholesteric structure.

After the thermal "training" [75,76] of the ds DNA CLCD, DAU solution was added to this dispersion (Figure 16, curve 2). It results in the appearance of an additional band located in the absorption region of DAU chromophores ($\lambda \sim 510$ nm). The amplitude of this band grows with the increase in DAU concentration and reaches an equilibrium value at such DAU concentration that corresponds to the maximal extent of the DNA saturation by DAU.

The negative sign of the band at $\lambda \sim 510$ nm shows that the orientation of DAU molecules coincide with the orientation of the nitrogen bases in respect to the long axis of the DNA molecules. It is possible if DAU molecules are intercalated between the nitrogen base pairs of ds DNA molecules in the content of the dispersion particles. Under these conditions the reactive groups of DAU (Figure 14b) are unavailable for chemical reactions [86].

To induce the formation of the "external" complexes of DAU with ds DNA (Figure 14c), the ds DNA CLCD was treated by DAU solutions, the concentrations of which exceed the equilibrium value. After that the solution of $CuCl_2$ was added. The addition of $CuCl_2$ leads to manifold increase of the band at $\lambda \sim 510$ nm (curves 3–5). This process takes about one hour (see inset in Figure 16). The amplification of the band in the visible region of the CD spectrum means that a new pattern of fixation of DAU nearby ds DNA molecules ordered in the quasinematic layers has appeared.

The multiple dilution of PEG-containing solution, in which the DNA CLCD was processed with DAU and $CuCl_2$, to conditions of full disintegration of the LC structure of the initial dispersion particles [27] does not lead to an appreciable decrease in the specific abnormal optical activity of the DNA CLCD (taking into account simply the decrease in the concentration of the DNA CLCD particles due to dilution). It means that the spatial orientation of adjacent ds DNA molecules in quasinematic layers of the CLCD treated by DAU and $CuCl_2$ is not violated even outside the "boundary" conditions [27].

Hence, the osmotic pressure of PEG-containing aqueous-salt solution is not the main factor affecting the pattern of packing ds DNA molecules in a CLCD particle.

The created structure can exist not only in solutions with a very low (up to zero) concentration of PEG but also in solutions with low ionic strength. It means, the adjacent ds DNA molecules in quasinematic layers of a single dispersion particles are cross-linked by (DAU-Cu^{2+}) complexes and these complexes stabilize the spatial structure of the CLCD particles.

The formation of cross-links between ds DNA molecules ordered in quasinematic layers of dispersion particle depends on the number of DAU molecules and on the number of copper atoms, as well as on the average distance between DNA molecules in the spatial structure of a dispersion particle [86].

The analysis of more than 10 DAU analogues, which differ by the presence and the position of substituents at the aglycon, has shown that the presence of reactive oxygen atoms in the 5, 6 and 11, 12 positions of anthracyclines is one of the essential conditions for the amplification of abnormal optical activity upon building of cross-links (chelate bridges). Consequently, each copper atom in the bridge forms 4 bonds with coplanar oxygen atoms [87]. This reflects the well-known stereochemical and electronic properties of chelate complexes in which Cu^{2+} ions interact with 4 oxygen atoms of anthracyclines [81,88–90].

If Cu^{2+} ion forms a chelate complex with four reactive oxygen atoms, this ion is in d^9 state, which exhibits a nonzero magnetic moment [91]. It allows one to directly estimate the number of Cu^{2+} ions in one cross-link (nanobridge) between adjacent ds DNA molecules ordered in quasinematic layers of one cholesteric liquid-crystalline particle by low-temperature magnetometry [91].

Taking into account the distance between ds DNA molecules in dispersion particles formed at $C_{PEG} = 170$ mg·mL^{-1}, the results obtained by low-temperature magnetometry and supported by the theoretical calculations [87,92–94] allow one to accept that the nanobridge between adjacent ds DNA molecules contains six Cu^{2+} ions. Besides, the theoretical evaluations, based on the analysis of the dependences of the amplitude of the band at λ ~ 510 nm on the number of DAU molecules (an isotropically oriented near ds DNA surface) and on the number of copper atoms [92,93] has shown that one nanobridge in dispersion particles (at C_{PEG} = 170 mg·mL^{-1}) can include from four to six DAU molecules and its final structure is shown in Figure 15c.

The nanobridges (cross-links) are planar chelate complexes. The stabilization of the nanobridges is achieved by the rigid fixation of the [-(Cu^{2+}-DAU ... - ... DAU-Cu^{2+})$_n$] chelate system between two adjacent DNA molecules. It means that the spatial final system of nanobridge looks like [DNA-extDAU-(Cu^{2+}-DAU ... - ... DAU-Cu^{2+})$_n$-extDAU-DNA].

The directions of the long axes of the planar nanobridges are perpendicular to the direction of the long axes of ds DNA molecules ordered in quasinematic layers of the dispersion particles, and, hence, to the cholesteric axis. In this case, the helical twist of the quasinematic layers means the spatial rotation of the "coloured" nanobridges and they are "visible" by means of CD spectroscopy. If this is the case, one can suppose that the amplitude of an abnormal band in the CD spectrum of cholesteric dispersion particles depends upon the length of nanobridges between adjacent ds DNA molecules, which, in turn, depends on the distance between these molecules.

The formation of the nanobridges (cross-links) between ds DNA molecules ordered in quasinematic layers of a single dispersion particle results in the creation of three-dimensional, "integrated" structure of the particle, which includes all ds DNA molecules as well as numerous (DAU-Cu^{2+}) bridges in the content of a single particle. This high molecular mass structure is incompatible with a PEG-containing solution. The stability of the "rigid" structure is determined by the number and the properties of the nanobridges rather than the properties of the initial PEG-containing solution. Hence, the transition of ds DNA particles from the "liquid-like" to the "rigid" (gel-like) state takes place [42].

Due to the presence of the nanobridges between adjacent ds DNA molecules, the spatial structure of "rigid" particles remains stable for a long time (months) at PEG and NaCl concentrations significantly below those at which the CLCD particles, not stabilized with nanobridges, can exist. Consequently, the created structure can exist not only in solutions with low ionic strength but also in solutions with a very low (up to zero) concentration of PEG, that is, under the conditions of a low osmotic pressure of the solution.

It is worth to noting that the rigid particle size can be different from the liquid-like particle size. Indeed although the intermolecular potential is overall repulsive, due to the nanobridges, it contains a local minimum at relatively short distances corresponding to the characteristic bridge length. Then the competition between the standard aggregation driven by osmotic pressure and bridge induced potential energy minimum, kinetics of macroscopic phase formation, may proceed non-monotonously in time. We defer the detailed investigation of this phenomenon for a future work.

Taking the results shown above into account, the possibility to investigate the properties of "rigid" particles of ds DNA by experimental techniques appears. For instance, one can immobilize "rigid" particles cross-linked by nanobridges onto the surface of nuclear membrane filter and directly directly visualize their size.

The AFM images of "rigid" particles formed at C_{PEG} = 170 mg·mL^{-1} have been recorded. The shape of the particles is close to spherocylinders. Estimation of the size of many particles demonstrates that the mean size of these particles is close to 500–600 nm, which is in good agreement with the data on particle size theoretically calculated for the initial PEG-containing solution with constant osmotic pressure [27].

Therefore, a few new features, that is, the appearance of an additional band at λ 510 nm at the treatment of the ds DNA dispersion by DAU, the formation of cross-links between adjacent ds DNA molecules in quasinematic layers and the creation of the "rigid" structure are distinctive of the dispersions with the cholesteric pattern of ds DNA molecules ordering.

6. The Optical Evidence of the Cholesteric Packing of ds DNA Molecules in a New Optically Active ("Re-Entrant" Cholesteric) Phase

The steps of the formation of the initial LCD particles with hexagonal packing of ds DNA molecules and their transformation to the optically active state are illustrated in Figure 17.

(Again, in the CD spectrum of the initial ds DNA LCD obtained as the result of the phase exclusion of ds DNA molecules from a PEG-containing solution (C_{PEG} = 280 mg·mL^{-1}) at room temperature there is a low intensity band of unknown origin. This non-specific band, which appears in the CD spectrum in the ds DNA absorption region (λ ~ 270 nm) confirms the formation of a dispersion, particles of which possess the hexagonal packing of ds DNA molecules [25]. Very low amplitude of this band is accepted as a zero value (see also Figure 8b, domain III).

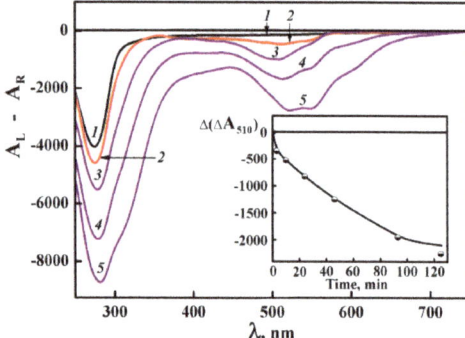

Figure 17. The CA spectra of the ds DNA LCD formed in an aqueous-salt PEG-containing solution (curve 1) and successively processing by DAU (curve 2) and CuCl$_2$ (curves 3–5) solutions. Curve 1—the ds DNA LCD after thermal "training". Curve 2—the ds DNA LCD treated by DAU (C_{DAU} = 5.99 × 10^{-5} M). Curves 3–5—the ds DNA LCD treated by DAU and CuCl$_2$ (C_{Cu} = 2.9 × 10^{-5} M) taken after 10 min, 46 min and 125 min. C_{PEG} = 280 mg·mL^{-1} (domain III), C_{DNA} = 10 μg·mL^{-1},0.3M NaCl + 0.002M Na$^+$-phosphate buffer.$(A_L - A_R) \times 10^{-6}$ optical units, 22 °C, L = 1 cm. Inset—the dependence of the amplitude of the band at 510 nm in the CD spectrum of ds DNA LCD treated by DAU and CuCl$_2$ upon time.

In addition, the results of an X-ray study of unoriented samples of phase, obtained by the sedimentation of LCD particles and our additional control of the ds DNA secondary structure, confirm that ds DNA molecules retain the B-form under these conditions. However, in the contrast to the previous case (Section 5) the mean distance between DNA molecules in the dispersion particles formed at C_{PEG} = 280 mg·mL^{-1} is about 2.5 nm.

Curve 1 in Figure 17 is the CD spectrum after the thermal "training" of the ds DNA dispersion, that is, after its heating to 80°C and subsequent cooling to the room temperature. One can see that the thermal "training" of this dispersion is accompanied by a remarkable optical effect, that is, an abnormal negative band arises in the CD spectrum at λ = 270 nm. An appearance of an abnormal band means that the heating of the dispersion particles with hexagonal packing of ds DNA molecules (and therefore, with no abnormal optical activity) results in a transition of these particles into a new optically active state, that is, a new optically active phase is formed. (We used term "ds DNA-III" to designate this phase). The transition, which looks like a phase transition, testifies of a change in the spatial structure of the ds DNA dispersion particles.

The comparison of curve 1 in Figure 17 to curve 1 in Figure 16 demonstrates their close similarity. It permits us to suppose that the initial hexagonal packing of quasinematic layers of ds DNA molecules in the particles is transformed into the cholesteric packing upon their heating.

However, the optical texture of a thin layer of ds DNA phase, obtained from the heated dispersion particles formed at C_{PEG} = 280 mg·mL^{-1}, does not demonstrate the specific characters typical of the classical cholesterics, in particular, we cannot see the fingerprint texture [95]. Since the distance between ds DNA molecules in a new optically active phase formed at C_{PEG} = 280 mg·mL^{-1} is small (2.5 nm), the simplest rationalization of the absence of the fingerprint texture typical of classical DNA cholesterics is that the pitch of the helical twist of the new structure of the dispersion particles formed at C_{PEG} = 280 mg·mL^{-1} is so small, that it cannot be detected by the polarization microscope.

To check the hypothesis about the cholesteric packing of ds DNA molecules in the new optically active phase, we used an approach considered above.

Fixed volumes of DAU solution have been added to the dispersion particles of the new optically active phase (Figure 17, curve 2). In full correspondence with the results presented in Figure 16, the addition of DAU to the ds DNA-III is accompanied by the appearance of a new band located in the absorption region of DAU chromophores ($\lambda \sim 510$ nm). The amplitude of this band grows with the increase in DAU concentration and reaches the equilibrium value at DAU concentration that corresponds to the maximal saturation of the ds DNA-III with DAU molecules (curve 2). The equilibrium value of this band does not change upon further growth of DAU concentration.

The negative sign of the band at $\lambda \sim 510$ nm shows that the orientation of DAU in respect to the axis of the DNA molecules in the content of ds DNA-III coincides with the orientation of nitrogen bases.

According to the results presented in Section 5, the appearance of an additional band in the CD spectrum in DAU absorption region is the first evidence in favour of the spatially twisted (cholesteric) structure of ds DNA-III phase.

We induced the formation of the "external" complexes of DAU with ds DNA-III particles by increase of DAU concentration. Then, the obtained mixture was treated with $CuCl_2$ solution. The addition of $CuCl_2$ solution to the ds DNA-III, treated by DAU, leads to manifold increase in the band amplitude at $\lambda \sim 510$ nm (Figure 17, curves 3–5). The amplification of this band is the second evidence in favour of the spatially twisted (cholesteric) structure of the ds DNA-III phase.

Hence, one can consider the appearance of the additional band at λ 510 nm at the treatment of ds DNA-III dispersion by DAU, its amplification at addition of $CuCl_2$ solution as the distinctive factors of the dispersions with the cholesteric ordering of ds DNA molecules. Based on the similarity between the abnormal optical properties of ds DNA-III phase and the standard ds DNA cholesterics, we designate this new state as the "re-entrant" cholesteric phase [77].

However, comparison of Figure 17 to Figure 16 attracts attention to following items: (i) the process of amplification of the band at $\lambda \sim 510$ nm (Figure 17, see inset) takes more time in comparison to the case shown in Figure 16, (ii) the maximal value of the band at $\lambda \sim 510$ in Figure 17 is about 3 times smaller in comparison to Figure 16. It means that the abnormal optical properties typical of two types of the cholesteric dispersion particles (formed at C_{PEG} 170 and 280 mg·mL^{-1}) are related, as minimum, to the distance between adjacent ds DNA molecules ordered in quasinematic layers. It is not excluded that the dense packing ds DNA molecules in dispersion particles formed at C_{PEG} 280 mg·mL^{-1} prevents the correct spatial "phasing" of adjacent ds DNA molecules in quasinematic layers and influences, probably, the pattern of DAU interaction with ds DNA molecules.

Taking into account the small distance between molecules in dispersion particles formed at C_{PEG} 280 mg·mL^{-1} (i.e., ds DNA-III), we can suppose that there are two reasons for the increase in the amplitude of the band in the CD spectrum of ds DNA-III dispersion treated by high DAU concentration and $CuCl_2$.

The first reason consists in the formation of a small size cross-links ("nanobridges") of type [-(Cu^{2+}-DAU ... - ... DAU-Cu^{2+})$_{n-}$] between two adjacent DNA molecules ordered in quasinematic layers of dispersion particle. In this case, the helical twist of the quasinematic layers means the spatial rotation of the "coloured" nanobridges and they are "visible" in the CD spectrum at $\lambda \sim 510$ nm (Figure 17). The formation of the nanobridges (cross-links) between ds DNA molecules ordered in quasinematic layers of a single dispersion particle results in the creation of the "integrated" structure of the particle. This structure must be incompatible with a PEG-containing solution and the transition of ds DNA-III particles from the "liquid-like" to the "rigid" (gel-like) state will takes place.

However, there is the other reason for the increase in the amplitude of band in the CD spectrum of ds DNA-**III** dispersion treated by high DAU concentration and CuCl$_2$ [87,95].

An intense band at $\lambda \sim 510$ nm may reflect the appearance of a new pattern of anisotropic arrangement of DAU molecules, which is realized, mainly, at small distances between the ds DNA molecules in dispersion particles. Owing to stacking interaction between DAU molecules forming the "external" complexes and free DAU molecules in solution, a "shell" of DAU is formed in proximity to the surface of ds DNA-**III** molecule, where a portion of DAU molecules is linked by copper ions (Figure 18). It should be noted that, in principle, the beginning of this "shell" may be not only a DAU molecule of "external" complexes but also a copper ion bound to ds DNA bases. At a small distance between molecules in the ds DNA-**III** dispersion particles, the question of the steric hindrances arising upon such an arrangement of adjacent DAU molecules, in particular, the location of sugar residues, remains open. The orientation of DAU molecules in the created structure may coincide with the orientation of nitrogen base pairs. It is obvious that in the case of the "shell" structure, DAU molecules are ordered in quasinematic layer. The helical twist of the quasinematic layers results in the spatial rotation of the created "shell" structure and it is "visible" in the CD spectrum. Whether there is in this case the transition from "liquid-like" to "rigid" (gel-like) state of ds DNA-**III** dispersion particle, remains unknown.

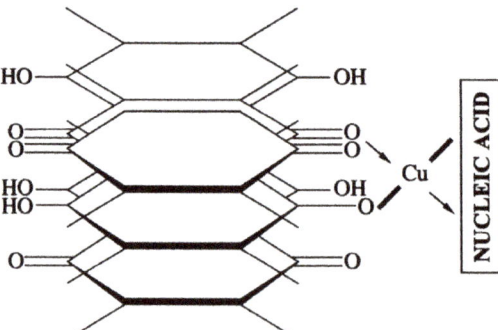

Figure 18. Hypothetical scheme of spatial ordering of DAU molecules in proximity to ds DNA surface on account of stacking interaction of these molecules (side view).

The both aforesaid arguments confirm the helical twist of the quasinematic layers of dispersion particles of ds DNA-**III**, that is, they strongly suggest the existence of the "re-entrant" cholesteric structure of ds DNA-**III**.

The investigation of the details of an appearance of an intense band at $\lambda \sim 510$ nm in the case of all dispersion particles formed in domain **III** and differ by the distances between ds DNA molecules in quasinematic layers is now in progress.

The results presented in this review demonstrate that "liquid-like" and "rigid" (solid) ds DNA dispersion particles have the unique physicochemical properties. This opens a gate for their application in various fields of science and technology, for instance:

(i) "Liquid-like" particles based on ds DNA molecules can be considered as polyfunctional sensing elements for optical analytical systems permitting the detection of biologically active compounds: antibiotics, genotoxicants and so forth, in laboratory and physiological liquids (application field: medicine, ecology and biotechnology);

(ii) "Rigid" particles, where the ds DNA concentration exceeding 200 mg·mL^{-1}, can be used as carriers of genetic material or various biologically-active or chemical substances introduced into these structures (application field: medicine and biotechnology).

7. Conclusions

The phase exclusion of ds linear, low molecular mass DNA molecules from aqueous-salt PEG-containing solutions at room temperature results in the formation of LCD particles with the cholesteric and the hexagonal packing mode of the molecules. However, the heating of dispersion particles with the hexagonal packing of ds DNA molecules results in a new phase transition accompanied by the appearance of an abnormal optical activity in the particles. Obtained results are explained in the framework of a concept of orientationally ordered "quasinematic" layers formed by ds DNA molecules, with a parallel alignment in the hexagonal structure. These layers can adopt a spatially twisted configuration at temperature increase; and as a result of this process, a new optically active structure is formed. This structure was termed the "re-entrant" cholesteric phase. This as one can expectour results indicate than the pattern of packing of the adjacent "quasinematic" layers of the ds DNA molecules in dispersion particles, obtained in PEG-containing aqueous-salt solutions, is determined not only by the osmotic pressure of the solution but also by its temperature.

To prove the cholesteric pattern of ds DNA molecules in a new optically active phase an approach based on the transition from the "liquid-like" to the "rigid" state of dispersion particles was used.

Author Contributions: Y.Y. and S.S. (Sergey Skuridin) conceived and designed the experiments; S.S. (Sergey Semenov) performed theoretical calculations of the circular dichroism spectra; V.S. and S.S. (Sergey Skuridin) performed the experiments; Y.Y., S.S. (Sergey Skuridin) and E.K. analysed the data; Y.Y. and E.K. wrote the paper with an input from all co-authors.

Funding: Yuri Yevdokimov, Sergey Skuridin, Viktor Salyanov were financially supported by the Russian Science Foundation (Grant number 16-15-00041).

Conflicts of Interest: The authors declare no conflict of interest.

References

1. Robinson, C. Liquid crystalline structures in polypeptides solutions. *Tetrahedron* **1961**, *13*, 219–234. [CrossRef]
2. Luzzati, V.; Nicolaieff, A. The structure of nucleohistones and nucleoprotamines. *J. Mol. Biol.* **1963**, *7*, 142–163. [CrossRef]
3. Iizuka, I. Some new finding in the liquid crystals of sodium salt of deoxyribonucleic acid. *Polym. J.* **1977**, *9*, 173–180. [CrossRef]
4. Senechal, E.; Maret, G.; Dransfeld, K. Long-range order of nucleic acids in aqueous solutions. *Int. J. Biol. Macromol.* **1980**, *2*, 256–262. [CrossRef]
5. Potaman, V.N.; Alexeev, D.G.; Skuratovskii, I.Y.; Rabinovich, A.Z.; Shlyakhtenko, L.S. Study of DNA films by CD, X-ray and polarization microscopy techniques. *Nucleic Acids Res.* **1981**, *9*, 55–64. [CrossRef] [PubMed]
6. Rill, R.L.; Hilliard, P.R.; Levy, G.C. Spontaneous ordering of DNA. *J. Biol. Chem.* **1983**, *258*, 250–256. [PubMed]
7. Brandes, R.; Kearns, D.R. Magnetic ordering of DNA liquid crystals. *Biochemistry* **1986**, *25*, 5890–5895. [CrossRef] [PubMed]
8. Strzelecka, T.E.; Rill, R.L. Solid-state phosphorus-31 NMR studies of DNA liquid crystalline phases. The isotropic to cholesteric transition. *J. Am. Chem. Soc.* **1987**, *109*, 4513–4518. [CrossRef]
9. Merchant, K.; Rill, R. Isotropic to anisotropic phase transition of extremely long DNA in an aqueous saline solution. *Macromolecules* **1994**, *27*, 2365–2370. [CrossRef]
10. Wissenburg, P.; Odjik, T.; Cirkel, P.; Mandel, M. Multimolecular aggregation of monoclonal DNA in concentrated isotropic solutions. *Macromolecules* **1995**, *28*, 2315–2328. [CrossRef]
11. Rau, D.C.; Lee, B.; Parsegian, V.A. Measurement of the repulsive force between polyelectrolyte molecules in ionic solution: Hydration forces between parallel DNA double helices. *Proc. Natl. Acad. Sci. USA* **1984**, *81*, 2621–2625. [CrossRef] [PubMed]
12. Parsegian, V.A.; Rand, R.P.; Fuller, N.L.; Rau, D.C. Osmotic stress for the direct measurement of intermolecular forces. *Methods Enzymol.* **1986**, *127*, 400–416. [CrossRef] [PubMed]
13. Rau, D.C.; Parsegian, V.A. Direct measurement of the intermolecular forces between counterion-condensed DNA double helices. Evidence for long range attractive hydration forces. *Biophys. J.* **1992**, *61*, 246–259. [CrossRef]

14. Todd, B.A.; Parsegian, V.A.; Shirahata, A.; Thomas, T.J.; Rau, D.C. Attractive forces between cation condensed DNA double helices. *Biophys. J.* **2008**, *94*, 4775–4782. [CrossRef]
15. Livolant, F.; Bouligand, Y. Liquid crystalline phases given by helical biological polymers (DNA, PBLG and xantan). Columnar textures. *J. Phys. (France)* **1986**, *47*, 1813–1827. [CrossRef]
16. Livolant, F.; Levelut, A.M.; Doucet, J.; Benoit, J.P. The highly concentrated liquid-crystalline phase of DNA is columnar hexagonal. *Nature* **1989**, *339*, 724–726. [CrossRef] [PubMed]
17. Rill, R.L.; Strzelecka, T.E.; Davidson, M.W.; van Winkle, D.H. Ordered phases in concentrated DNA solutions. *Physica A* **1991**, *176*, 87–116. [CrossRef]
18. Kassapidou, K.; Jesse, W.; van Dijk, J.F.; van der Maarel, J.R. Liquid crystal formation in DNA fragment solutions. *Biopolymers* **1998**, *46*, 31–37. [CrossRef]
19. Livolant, F. Cholesteric organization of DNA in vivo and in vitro. *Eur. J. Cell Biol.* **1984**, *33*, 300–311.
20. Rill, R.L.; Livolant, F.; Aldrich, H.C.; Davidson, M.A. Electron microscopy of liquid crystalline DNA: Direct evidence of cholesteric-like organization of DNA in dinoflagellate chromosome. *Chromosoma* **1989**, *98*, 280–286. [CrossRef]
21. Livolant, F. Supramolecular organization of double-stranded DNA molecules in the columnar hexagonal liquid crystalline phase. An electron microscopic analysis using freeze-fracture methods. *J. Mol. Biol.* **1991**, *218*, 165–181. [CrossRef]
22. Leforestier, A.; Livolant, F. Cholesteric liquid crystalline DNA; a comparative analysis of cryofixation methods. *Biol. Cell* **1991**, *71*, 115–122. [CrossRef]
23. Durand, D.; Doucet, J.; Livolant, F. A study of the structure of highly concentrated phases of DNA by X-ray diffraction. *J. Phys. II France* **1992**, *2*, 1769–1783. [CrossRef]
24. Leforestier, A.; Livolant, F. Supramolecular ordering of DNA in the cholesteric liquid crystalline phase: An ultrastructural study. *Biophys. J.* **1993**, *65*, 56–72. [CrossRef]
25. Livolant, F.; Leforestier, A. Condensed phases of DNA: Structures and phase transitions. *Prog. Polym. Sci.* **1996**, *21*, 1115–1164. [CrossRef]
26. Mitov, M. Cholesteric liquid crystals in living matter. *Soft Matter* **2017**, *13*, 4176–4209. [CrossRef]
27. Yevdokimov, Y.M.; Salyanov, V.I.; Semenov, S.V.; Skuridin, S.G. *DNA Liquid Crystalline Dispersions and Nanoconstructions*; CRC Press (Taylor & Francis Group): Boca Raton, FL, USA, 2012; 258p.
28. Bouligand, Y. Liquid crystalline order in biological materials. In *Solid State Physics*; Liebert, L., Ed.; Academic: New York, NY, USA, 1978; (Suppl. 14), pp. 259–294.
29. Kamien, R.D.; Selinger, J.V. Order and frustration in chiral liquid crystals. *J. Phys.Condens. Matter* **2001**, *13*, R1. [CrossRef]
30. Lerman, L.S. A transition to a compact form of DNA in polymer solutions. *Proc. Natl. Acad. Sci. USA* **1971**, *68*, 1886–1890. [CrossRef] [PubMed]
31. Evdokimov, Y.M.; Platonov, A.L.; Tikhonenko, A.S.; Varshavsky, Y.M. A compact form of double-stranded DNA in solution. *FEBS Lett.* **1972**, *23*, 180–184. [CrossRef]
32. Akimenko, N.M.; Dijakova, E.B.; Evdokimov, Y.M.; Frisman, E.V.; Varshavsky, Y.M. Viscosimetric study of compact form of DNA in water-salt solutions containing polyethyleneglycol. *FEBS Lett.* **1973**, *38*, 61–63. [CrossRef]
33. Maniatis, T.; Venable, J.H.; Lerman, L.S. The structure of Ψ DNA. *J. Mol. Biol.* **1974**, *84*, 37–64. [CrossRef]
34. Lerman, L.S. Chromosomal analogues: Long-range order in Ψ -condensed DNA. *Cold Spring Harb. Symp. Quant. Biol.* **1974**, *38*, 59–73.
35. Poglazov, B.F.; Tikhonenko, A.S.; Engelhardt, V.A. ATP action on DNA release from bacteriophage. *Proc. USSR Acad. Sci.* **1962**, *145*, 450–452.
36. Laemmli, U.K. Characterization of DNA condensates induced by poly(ethylene oxide) and polylysine. *Proc. Natl. Acad. Sci. USA* **1975**, *72*, 4288–4292. [CrossRef]
37. Hud, N.V.; Vilfan, I.D. Toroidal DNA condensates: Unraveling the fine structure and the role of nucleation in determining size. *Annu. Rev. Biophys. Biomol. Struct.* **2005**, *34*, 295–318. [CrossRef] [PubMed]
38. Hoang, T.X.; Giacometti, A.; Podgornik, R.; Nguyen, N.T.; Banavar, J.R.; Maritan, A. From toroidal to rod-like condensates of semiflexible polymers. *J. Chem. Phys.* **2014**, *140*, 064902. [CrossRef] [PubMed]
39. Yevdokimov, Y.M.; Skuridin, S.G.; Lortkipanidze, G.B. Liquid-crystalline dispersions of nucleic acids. *Liq. Cryst.* **1992**, *12*, 1–16. [CrossRef]

40. Goldar, A.; Thomson, H.; Seddon, J.M. Structure of DNA cholesteric spherulitic droplet dispersions. *J. Phys. Condens. Matter* **2007**, *20*, 9. [CrossRef]
41. Biswas, N.; Ichikawa, M.; Datta, A.; Sato, Y.T.; Yanagisawa, M.; Yoshikawa, K. Phase separation in crowded micro-spheroids: DNA-PEG system. *Chem. Phys. Lett.* **2012**, *539–540*, 157–162. [CrossRef]
42. Yevdokimov, Y.M.; Skuridin, S.G.; Salyanov, V.I.; Bykov, V.A.; Palumbo, M. Structural DNA nanothechnology: Liquid-crystalline approach. In *Biotechnology.Applied Synthetic Biology*; Singh, V., Ed.; Studium Press LLC: Houston, TX, USA, 2014; Volume 4, pp. 327–380.
43. Yevdokimov, Y.M.; Salyanov, V.I.; Savateev, M.N.; Dubinskaya, V.A.; Skuridin, S.G. Analysis of "solid" nanoconstructs formed of liquid-crystalline DNA dispersion particles by the method of atomic force microscopy. *Technol. Living Syst.* **2013**, *10*, 20–27. (In Russian)
44. Livolant, F. Ordered phases of DNA in vivo and in vitro. *Phys. A* **1991**, *176*, 117–137. [CrossRef]
45. Adamczyk, A. Phase transition in freely suspended smectic droplets. Cotton-Mouton technique, architecture of droplets and formation nematoids. *Mol. Cryst. Liq. Cryst.* **1989**, *170*, 53–69. [CrossRef]
46. Chiccoli, C.; Pasini, P.; Semeria, F.; Zannoni, C. Computer simulations of nematic droplets with toroidal boundary conditions. *Mol. Cryst. Liq. Cryst.* **1992**, *221*, 19–28. [CrossRef]
47. Chandrasekhar, S. *Liquid Crystals*, 2nd ed.; Cambridge University Press: Cambridge, UK, 1992; p. 480.
48. Papkov, S.P.; Kulichikhin, V.G. *The Liquid Crystal State of Polymers*; Khimiya: Moscow, Russia, 1977; 240p. (In Russian)
49. Chilaya, G.S.; Lisetskii, L.N. Helical twist in cholesteric mesophases. *Sov. Phys. Usp.* **1981**, *24*, 496–510. [CrossRef]
50. Kornyshev, A.A.; Leikin, S.; Malinin, S.V. Chiral electrostatic interaction and cholesteric liquid crystals of DNA. *Eur. Phys. J. E* **2002**, *7*, 83–93. [CrossRef]
51. Yevdokimov, Y.M.; Salyanov, V.I.; Skuridin, S.G.; Semenov, S.V.; Kompanets, O.N. The CD Spectra of Double-Stranded DNA Liquid Crystalline Dispersions. In *Circular Dichroism: Theory and Spectroscopy*; Rogders, D.S., Ed.; Nova Science Publishers: New York, NY, USA, 2012; pp. 5–75.
52. Semenov, S.V.; Yevdokimov, Y.M. Circular dichroism of DNA liquid-crystalline dispersion particles. *Biophysics* **2015**, *60*, 188–196. [CrossRef]
53. Keller, D.; Bustamante, C. Theory of the interaction of light with large inhomogeneous molecular aggregates. I. Absorption. *J. Chem. Phys.* **1986**, *84*, 2961–2971. [CrossRef]
54. Keller, D.; Bustamante, C. Theory of the interaction of light with large inhomogeneous molecular aggregates. II. Psi-type circular dichroism. *J. Chem. Phys.* **1986**, *84*, 2972–2980. [CrossRef]
55. Kim, M.-H.; Ulibarri, L.; Keller, D.; Maestre, M.F.; Bustamante, C. The psi-type circular dichroism of large molecular aggregates. III. Calculations. *J. Chem. Phys.* **1986**, *84*, 2981–2989. [CrossRef]
56. Belyakov, V.A.; Sonin, A.S. *Optics of Cholesteric Liquid Crystals*; Nauka: Moscow, Russia, 1982; 360p. (In Russian)
57. Belyakov, V.A.; Dmitrienko, V.E. *Optics of Chiral Liquid Crystals (Sov. Sci. Rev. Sect. A)*; Routledge: Abingdon-on-Thames, UK, 1989; 222p.
58. Saeva, F.D.; Wysocki, J.J. Induced circular dichroism in cholesteric liquid crystals. *J. Am. Chem. Soc.* **1971**, *93*, 5928–5929. [CrossRef]
59. Sackman, E.; Voss, J. Circular dichroism of helically arranged molecules in cholesteric phases. *Chem. Phys. Lett.* **1972**, *14*, 528–532. [CrossRef]
60. De Vries, H. Rotatory power and other optical properties of certain liquid crystals. *Acta Cryst.* **1951**, *4*, 219–226. [CrossRef]
61. Mauguin, C. Sur les cristaux liquids de Lehman. *Bull. Soc. Franc. Minéral.* **1911**, *34*, 71–117.
62. Nordén, B. Application of linear dichroism spectroscopy. *Appl.Spectrosc. Rev.* **1978**, *14*, 157–248. [CrossRef]
63. Belyakov, V.A.; Orlov, V.P.; Semenov, S.V.; Skuridin, S.G.; Yevdokimov, Y.M. Comparison of calculated and observed CD spectra of liquid crystalline dispersions formed from double-stranded DNA and from DNA complexes with coloured compounds. *Liq. Cryst.* **1996**, *20*, 777–784. [CrossRef]
64. Yevdokimov, Y.M.; Salyanov, V.I.; Skuridin, S.G.; Semenov, S.V.; Kompanets, O.N. *The CD Spectra of Double-Stranded DNA Liquid Crystalline Dispersions*; Nova Science Publishers: New York, NY, USA, 2011; 103p.
65. Harris, A.B.; Kamien, R.D.; Lubensky, T.C. Microscopic origin of cholesteric pitch. *Phys. Rev. Lett.* **1997**, *78*, 1476–1479. [CrossRef]

66. Harris, A.B.; Kamien, R.D.; Lubensky, T.C. Molecular chirality and chiral parameters. *Rev. Modern Phys.* **1999**, *71*, 1745–1757. [CrossRef]
67. Bloomfield, V.A.; Crothers, D.M.; Tinoko, I. *Physical Chemistry of Nucleic Acids*; Harper & Row: New York, NY, USA, 1974; 517p.
68. Evdokimov, Y.M.; Pyatigopckaya, T.L.; Polivtsev, O.F.; Akimenko, N.M.; Tsvankin, D.Y.; Varshavsky, Y.M. DNA compact form. 8. X-ray diffraction study of DNA compact particles, formed in solutions, containing poly (ethylene glycol). *Mol. Biol.* **1976**, *10*, 1221–1229. (In Russian)
69. Skuridin, S.G.; Damaschun, H.; Damaschun, G.; Yevdokimov, Y.M.; Misselwitz, R. Polymer condensed DNA: A study by small-angle X-ray scattering, intermediate-angle X-ray scattering and cicular dichroitic spectroscopy. *Stud. Biophys.* **1986**, *112*, 139–150.
70. Yevdokimov, Y.M.; Skuridin, S.G.; Semenov, S.V.; Dadinova, L.A.; Salyanov, V.I.; Kats, E.I. Re-entrant cholesteric phase in DNA liquid-crystalline dispersion particles. *J. Biol. Phys.* **2017**, *43*, 45–68. [CrossRef] [PubMed]
71. Lis, J.T.; Schleif, R.R. Size fractionation of double-stranded DNA by precipitation with polyethylene glycol. *Nucleic Acids Res.* **1975**, *2*, 383–390. [CrossRef]
72. Brunner, W.C.; Maestre, M.F. Circular dichroism of films of polynucleotides. *Biopolymers* **1974**, *13*, 345–357. [CrossRef] [PubMed]
73. Yevdokimov, Y.M.; Skuridin, S.G.; Salyanov, V.I. The liquid-crystalline phases of double-stranded nucleic acids in vitro and in vivo. *Liq. Cryst.* **1988**, *3*, 1443–1459. [CrossRef]
74. Yevdokimov, Y.M.; Skuridin, S.G.; Salyanov, V.I.; Muzipov, E.R.; Semenov, S.V.; Kats, E.I. Double-stranded DNA packing in particles of liquid-crystalline dispersions and liquid-crystalline phases obtained from these particles. *Liq. Cryst. Their Appl.* **2018**, *18*, 64–85. [CrossRef]
75. Sonin, A.S. *Introduction to the Physics of Liquid Crystals*; Nauka: Moscow, Russia, 1983; 320p. (In Russian)
76. Sundaresan, N.; Thomas, T.; Thomas, T.J.; Pillai, C.K.S. Lithium ion induced stabilization of liquid crystalline DNA. *Macromol. Biosci.* **2006**, *6*, 27–32. [CrossRef] [PubMed]
77. Yevdokimov, Y.M.; Skuridin, S.G.; Salyanov, V.I.; Semenov, S.V.; Shtykova, E.V.; Dadinova, L.A.; Kompanets, O.N.; Kats, E.I. The re-entrant cholesteric phase of DNA. *Opt. Spectrosc.* **2017**, *123*, 56–69. [CrossRef]
78. Yevdokimov, Y.M.; Salyanov, V.I.; Skuridin, S.G.; Shtykova, E.V.; Khlebtsov, N.G.; Kats, E.I. Physicochemical and nanotechnological approaches to the design of "rigid" spatial structures of DNA. *Rus. Chem. Rev.* **2015**, *84*, 27–42. [CrossRef]
79. Doskocil, J.; Fric, I. Complex formation of daunomycin with double-stranded RNA. *FEBS Lett.* **1973**, *37*, 55–58. [CrossRef]
80. Barthelemy-Clavey, V.; Maurizot, J.-C.; Sicard, P.J. Etude spectrophotometrique du complexe DNA-daunorubicine. *Biochimie* **1973**, *55*, 859–868. [CrossRef]
81. Wells, A.F. Cooper, Silver and Gold. In *Structural Inorganic Chemistry*, 4th ed.; Oxford University Press: Oxford, UK, 1975; pp. 875–910.
82. Bersuker, I.B. *The Jahn-Teller Effect and Vibronic Interactions in Modern Chemistry*; Nauka: Moscow, Russia, 1987; 251p. (In Russian)
83. Basolo, F.; Pearson, R.G. *Mechanisms of Inorganic Reactions: A Study of Metal Complexes in Solution*, 2nd ed.; Wiley: New York, NY, USA, 1967; 701p.
84. Kaneko, M.; Tsuchida, E. Formation, characterization and catalytic activity of polymer-metal complexes. *J. Polym. Sci. Macromol. Rev.* **1981**, *16*, 397–522. [CrossRef]
85. Coble, H.D.; Holtzclaw, H.F. Chelate polymers of copper (II) various dihydroxyquinoid ligands. *J. Inorg. Nuclear Chem.* **1974**, *36*, 1049–1053. [CrossRef]
86. Yevdokimov, Y.M.; Skuridin, S.G.; Salyanov, V.I.; Bobrov, Y.A.; Bucharsky, V.A.; Kats, E.I. New optical evidence of the cholesteric packing of DNA molecules in "re-entrant" phase. *Chem. Phys. Lett.* **2019**, *717*, 59–68. [CrossRef]
87. Yevdokimov, Y.M.; Salyanov, V.I.; Nechipurenko, Y.D.; Skuridin, S.G.; Zakharov, M.A.; Spener, F.; Palumbo, M. Molecular constructions (superstructures) with adjustable properties based on double-stranded nucleic acids. *Mol. Biol.* **2003**, *37*, 293–306. [CrossRef]
88. Greenaway, F.T.; Dabrowiak, J.C. The binding of copper ions to daunomycin and adriamycin. *J. Inorg. Biochem.* **1982**, *16*, 91–107. [CrossRef]

89. Fishman, M.M.; Schwastz, I. Effect of divalent cations on the daunomycin-deoxyribonucleic acid complex. *Biochem. Pharmacol.* **1974**, *23*, 2147–2154. [CrossRef]
90. Roy, S.; Mondal, P.; Sengupta, P.S.; Dhak, D.; Santra, R.C.; Das, S.; Guin, P.S. Spectroscopic, computational and electrochemical studies on the formation of the copper complex of 1-amino-4-hydroxy-9,10-anthraquinone and effect of it on superoxide formation by NADH dehydrogenase. *Dalton Trans.* **2015**, *44*, 5428–5440. [CrossRef]
91. Nikiforov, V.N.; Kuznetsov, V.D.; Nechipurenko, Y.D.; Salyanov, V.I.; Yevdokimov, Y.M. Magnetic properties of copper as a constituent of nanobridges formed between spatially fixed deoxyribonucleic acid molecules. *JETP Lett.* **2005**, *81*, 264–266. [CrossRef]
92. Nechipurenko, Y.D.; Strel'tsov, S.A.; Yevdokimov, Y.M. Thermodynamic model of bridging between nucleic acid molecules in liquid crystal. *Biophysics* **2001**, *46*, 428–435.
93. Nechipurenko, Y.D.; Ryabokon, V.F.; Semenov, S.V.; Yevdokimov, Y.M. Thermodynamic models describing the formation of "bridges" between of nucleic acid molecules and liquid crystals. *Biophysics* **2003**, *48*, 594–601.
94. Ryabokon, V.F.; Nechipurenko, Y.D.; Semenov, S.V.; Yevdokimov, Y.M. "Bridges" between nucleic acid molecules in liquid crystals. Description from the point of view of the adsorption theory. *Liq. Cryst. Their Appl.* **2003**, *3*, 69–78. (In Russian)
95. Yevdokimov, Y.M.; Skuridin, S.G.; Salyanov, V.I.; Kats, E.I. Anomalous behavior of the DNA liquid-crystalline dispersion particles and their phases. *Chem. Phys. Lett.* **2018**, *707*, 154–159. [CrossRef]

© 2019 by the authors. Licensee MDPI, Basel, Switzerland. This article is an open access article distributed under the terms and conditions of the Creative Commons Attribution (CC BY) license (http://creativecommons.org/licenses/by/4.0/).

MDPI
St. Alban-Anlage 66
4052 Basel
Switzerland
Tel. +41 61 683 77 34
Fax +41 61 302 89 18
www.mdpi.com

Crystals Editorial Office
E-mail: crystals@mdpi.com
www.mdpi.com/journal/crystals

www.ingramcontent.com/pod-product-compliance
Lightning Source LLC
LaVergne TN
LVHW071959080526
838202LV00064B/6787